Hybrid Genetic Optimization for IC Chips Thermal Control

Advances in Metaheuristics

Series Editors:
Patrick Siarry, *Universite Paris-Est Creteil, France*
Anand J. Kulkarni, *Symbiosis Center for Research and Innovation, Pune, India*

Handbook of AI-based Metaheuristics
Edited by Patrick Siarry and Anand J. Kulkarni

Metaheuristic Algorithms in Industry 4.0
Edited by Pritesh Shah, Ravi Sekhar, Anand J. Kulkarni, Patrick Siarry

Constraint Handling in Cohort Intelligence Algorithm
Ishaan R. Kale, Anand J. Kulkarni

Hybrid Genetic Optimization for IC Chip Thermal Control: with MATLAB® applications
Mathew V. K., Tapano Kumar Hotta

For more information about this series please visit: https://www.routledge.com/Advances-in-Metaheuristics/book-series/AIM

Hybrid Genetic Optimization for IC Chips Thermal Control

with MATLAB® Applications

Mathew V. K.

Tapano Kumar Hotta

CRC Press
Taylor & Francis Group
Boca Raton London New York

CRC Press is an imprint of the
Taylor & Francis Group, an **informa** business

A CHAPMAN & HALL BOOK

MATLAB® is a trademark of The MathWorks, Inc. and is used with permission. The MathWorks does not warrant the accuracy of the text or exercises in this book. This book's use or discussion of MATLAB® software or related products does not constitute endorsement or sponsorship by The MathWorks of a particular pedagogical approach or particular use of the MATLAB® software.

First Edition published 2022
by CRC Press
6000 Broken Sound Parkway NW, Suite 300, Boca Raton, FL 33487-2742

and by CRC Press
4 Park Square, Milton Park, Abingdon, Oxon, OX14 4RN

CRC Press is an imprint of Taylor & Francis Group, LLC

© 2022 Mathew V. K., Tapano Kumar Hotta

ISBN: 978-1-032-03353-2 (hbk)
ISBN: 978-1-032-03685-4 (pbk)
ISBN: 978-1-003-18850-6 (ebk)

DOI: 10.1201/9781003188506

Typeset in Minion
by KnowledgeWorks Global Ltd.

Contents

About the Authors

Dr. Mathew V. K. holds a Ph.D. degree in Thermal Management Systems from Vellore Institute of Technology, Vellore, India and M. Tech. in Heat Power from the University of Pune, India. He is currently working as Research Coordinator at MIT-ADT University, Pune, India. He is proficient in experimental techniques and currently working in the field of heat transfer enhancement using active and passive cooling. His research interests include active and passive safety, crash energy management, battery thermal management system, computational fluid dynamics, heat transfer, numerical methods, and optimization using different algorithms (genetic algorithm, artificial neural network). Dr. Mathew is a Guest Editor in *Frontiers in Mechanical Engineering* for a research topic on "Performance Analysis on Heat Transfer Enhancement Techniques".

Dr. Tapano Kumar Hotta is currently working as Associate Professor in the School of Mechanical Engineering, VIT Vellore, India. He holds a Ph.D. degree in Mechanical Engineering from IIT Madras in the area of electronic cooling. His area of research in a broad sense includes active and passive cooling of electronic devices, heat transfer enhancement, optimization of thermal systems, thermal energy storage, etc. Dr. Hotta has about 13 years of teaching-cum-research experience and has around 40 publications to his credit in journals and conferences of international repute. He has guided more than 30 undergraduates, a dozen postgraduates, and 2 doctorate students for their project work. He has also published 2 patents. Dr. Hotta is a Guest Editor in *Frontiers in Mechanical Engineering* for a research topic on "Performance Analysis on Heat Transfer Enhancement Techniques". He is a member of the editorial board and reviewer for various international journals and conferences related to heat transfer. He has several awards to his credit; Silver Medal for being the topper in M-Tech Thermal Engineering, Research Award at VIT Vellore for the significant contribution to research, and Best Scientist Award (2021) in Mechanical Engineering from the IMRF foundation.

Preface

The continuous miniaturization of the integrated circuit (IC) chips (heat sources) and the increase in the sleekness (design) of the electronic components have led to the monumental rise of the volumetric heat generation in the electronic components. The book focuses on the detailed optimization strategy carried out to enhance the performance (temperature control) of the IC chips oriented at different positions on a switched-mode power supply (SMPS) board and cooled using air under various heat transfer modes. Seven asymmetric protruding IC chips mounted at different positions on an SMPS board are considered in the present study that is supplied with non-uniform heat fluxes.

The book has seven chapters. A brief description of the contents of each chapter is mentioned next.

Chapter 1 gives the introduction to electronic cooling, and the various techniques available for the cooling of the electronic components. The significance of the phase change material in the electronic cooling application along with the role of optimization in the heat transfer systems is also highlighted in this chapter.

Chapter 2 discusses the state of the art pertinent to the studies related to the cooling of the IC chips. The chapter identifies the scope for development towards the optimization studies related to the IC chip's cooling.

Chapter 3 discusses the design and selection of both the IC chips and the SMPS board. This chapter also presents the details about the different types of equipment and the methodology used for conducting the experiments under the various heat transfer modes using both air and PCM.

Chapter 4 proposes a methodology to decide the optimal configuration of seven asymmetric IC chips oriented at different positions on an SMPS board cooled under the mixed convection heat transfer mode using a numerical data-driven combined artificial neural network (ANN) and genetic algorithm (GA) technique.

Chapter 5 highlights the experimental and numerical investigations carried out under the forced convection heat transfer mode to study the substrate board orientation effect on the cooling of the IC chips. An experimental data-driven combined ANN-GA-based technique is employed to determine their optimal configuration.

Chapter 6 presents the experimental and numerical investigations of the Paraffin wax-based mini-channels placed adjacent to the IC chips. The experiments are carried out by filling paraffin wax inside the mini-channels for four different volumetric heat generation values.

Chapter 7 summarizes all the chapters of the book and presents the key conclusions of the present study. The scope for future research is also highlighted in this chapter.

Acknowledgement

I owe a debt of gratitude for the credible blessings of the almighty **Lord Jesus Christ** for giving me an opportunity to write the book in domain of Thermal Management.

I dedicate this book to my parents, **Shri Vijay Kumar and Smt. Shobha V.K.**, to make their dream come true and all their hardships to make me capable of conducting the research work. I also thank Dr. Anand Kulkarni for believing in me in writing this book.

The word "thank you" won't suffice to **Dr. Archana Chandak** who had made this research work conversion to a book from impossible to possible and dream come true. Her steadfast determination, for believing in me long after I'd lost belief in myself, and for sharing my wish to reach the goal of completing this task and always motivating me to keep a balance between research work and personal commitments. She has stood by me through all my travails, my absences, and my fits of pique and impatience. She gave me support and help, discussed ideas, and prevented several wrong turns.

I specially thank my son, **Christiano**, for understanding my workload and supporting me during my entire tenure and for all his sacrifices at such a tender age.

Mathew V. K.

Tapano Kumar Hotta is thankful for the support of his workplace Vellore Institute of Technology, Vellore, for extending help in all the possible ways towards the smooth execution of this book.

Tapano Kumar Hotta

Nomenclature

A	IC chip area, m^2
C	specific heat, J/kgK
D	ratio of the centroid of the IC chip with maximum temperature to the centroid of the channel or channel combination
D$_h$	channel hydraulic diameter, m
F$_o$	Fourier number, $\alpha t/L_c^2$
g	acceleration due to gravity, 9.81 m/s^2
Gr	Grashof number, $g\beta(T_{max} - T_\infty)H^3/\nu^3$
H	enclosure height, m
h	convective heat transfer coefficient, W/m^{2}K
k	thermal conductivity, W/m K
l$_c$	IC chip length, m
l	working area dimension of the substrate board, m
L$_c$	characteristic length of the IC chip, $4A/P_c$, m
m	mass of the PCM, kg
Nu	Nusselt number, hL/k_f
P	pressure, Pa
P$_c$	perimeter of the IC chip, m
Pe	non-dimensional parameter, D_h/D
Q	heat stored by the PCM, J
q	heat flux, W/m^2
Re	Reynolds number, $V L/\nu$
Ri	Richardson number, Gr/Re^2
T	temperature, K
t	time, sec
t$_c$	thickness of the IC chip, m
u, v, w	velocity component along x-, y-, and z-directions, respectively, m/s

V	velocity of air, m/s
w$_c$	width of the IC chip, m
X, Y	centroid, m
Σd_i^2	sum of the square of the distances for each configuration, from the centroid of each IC chip of that configuration to the centroid of that particular configuration, $\Sigma\,(X_i - X_c) + (Y_i - Y_c)$, m^2

GREEK SYMBOLS

ΔT_{ref}	reference temperature, qL/K$_s$, K
θ	non-dimensional temperature, $(T_{max} - T_\infty)/\Delta T_{ref}$
α	thermal diffusivity, m^2/s
λ	non-dimensional geometric distance parameter $\dfrac{\sum\limits_{i=1}^{7} d_i^2}{l^2 + Y_c^2}$
β	isobaric thermal expansion coefficient of fluids, 1/T$_{mean}$, 1/K
ρ	density of air, kg/m^3
ν	kinematic viscosity of air, m^2/s

SUBSCRIPTS

∞	ambient
c	configuration
corr	correlation
f	fluid
i	initial
m	melting
max	maximum
ref	reference
s	substrate board
sim	simulation

ABBREVIATIONS

ANN	Artificial neural network
GA	Genetic algorithm
IC	Integrated circuit
MRE	Mean relative error
MSE	Mean square error
PCM	Phase change material
RMS	Root mean square
SMPS	Switched-mode power supply

CHAPTER 1

Introduction to Electronic Cooling

1.1 INTRODUCTION

Thermal management of electronics is a key research area in the field of electronics and has fascinated many researchers in recent years. The issues related to the cooling of electronic components, different techniques available for the same, and the role of optimization in identifying the solutions to the heat transfer problems are discussed in this Chapter 1.

1.2 NEED FOR ELECTRONIC COOLING

The increasing heat flux demand from the electronic components is posing a greater challenge to the researchers. Ultimately their temperature shoots up. Hence, the role of the thermal management technique is extremely crucial to improve the reliability and performance of the electronic components. Figure 1.1 depicts the different factors affecting the failure of an electronic component, where temperature contributes to the highest factor of 55%.

The integrated circuit (IC) chips that are placed on the printed circuit board (PCB) are very sensitive to temperature as they are made of silicon. According to Arrhenius' law of failure rate, for every 20°C rise in the chip temperature, their breakdown rate is doubled. So, a good thermal design is highly essential for increasing the life span of the IC chips, and it is also imperative to decide their optimal arrangement on a switched-mode power supply (SMPS) board to enhance their cooling rate. Sections 1.3 and 1.4 highlight the available cooling techniques for the electronic components

DOI: 10.1201/9781003188506-1

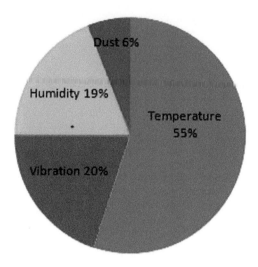

FIGURE 1.1 Factors affecting the failure of electronic components. (Murshed, 2016.)

and the role of optimization in determining the solution for the heat transfer problems, respectively.

1.3 PRINTED CIRCUIT BOARD AND INTEGRATED CIRCUIT CHIPS

A typically PCB, generally known as the substrate board (shown in Figure 1.2), is an electronic board that connects different circuit components. The boards are inexpensive, highly reliable, and compact. The PCBs are made of alternating layers of conducting and insulating materials. In usual practice, these are made of a low thermal conductivity material FR4 (reinforced epoxy laminate sheets, k = 1.35 W/mK). The conducting layers of the board are made of copper foil. The board is

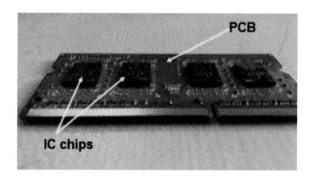

FIGURE 1.2 Photograph of 4GB RAM of a laptop. (Ref: Google image.)

coated with a solder mask that is usually green in colour; however, blue and red are the other available colours used for coating the PCB. The unwanted copper foil is then removed from the PCB after etching and then left with the desired copper tracers (wiring of the PCB). The different electronic components are soldered to the pads on the PCB, which are then connected to the tracers. The component size, their mounting on the PCB, and the heat dissipation rate from the PCB are some of the key parameters to be highlighted for the PCB design. For the present analysis, the PCB material is mimicked by a low thermal conductivity material, Bakelite (k = 0.24 W/mK).

An IC is a set of electronic circuits of semiconductor materials usually made of silicon. There is a wide application of the IC chips ranging from computers and mobile phones to home appliances. For the present study, the IC chips are referred to as discrete heat sources and are mimicked by aluminium. The IC chips are the heat-generating elements placed on the computer motherboards or the PCB of the electronic components. The asymmetric rectangular IC chips are considered for the present study, whose sizes are fetched from the data-sheet of the manufacturing catalogue.

1.4 VARIOUS COOLING TECHNIQUES

The cooling of the electronic components is achieved using several techniques, as given in Figure 1.3. The cooling techniques are broadly categorized into active and passive and are implemented depending upon the application.

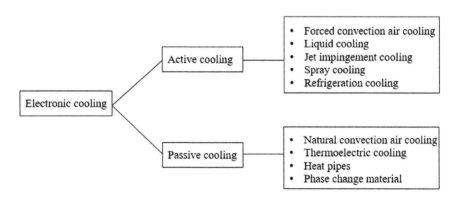

FIGURE 1.3 Different techniques available for the cooling of the electronic components.

Mechanically assisted cooling systems provide active cooling, as they require external energy for dissipating the heat from the electronic components and need a fan or blower for the cooling of these components. They offer high cooling capacity and allow the temperature control of the electronic components with the possibility of achieving it below the ambient temperature. The sub-types available under the active cooling techniques are forced convection cooling using air, jet impingement cooling, spray cooling, refrigeration cooling, liquid cooling, etc.

Nevertheless, the passive cooling techniques are also equally important, especially in the module-level thermal management and for the low heat flux removal rates. They do not need any external power sources or any external energy for dissipating the heat from the electronic components. The cooling rate of the components is achieved naturally using the heat sinks integrated with the fins. The sub-types under these techniques are natural convection air cooling, thermoelectric cooling, heat pipe-based cooling, phase change material (PCM)-based cooling, etc.

However, these conventional cooling methods cannot deal with the increasing heat flux demand rate of the electronic components. Hence, different innovative cooling techniques have been implemented for enhancing the heat removal rate from the components to maintain their temperature within the safe limit ($\leq 100°C$). The electronics industries are finding a stiff challenge for removing high heat fluxes (around 300 W/cm^2) while maintaining the component temperature below 100°C. Due to the miniaturization of the electronic components, the effective area available for heat dissipation is getting reduced, which leads to the development of emerging domains like mini-channels and micro-channels for the effective cooling of the electronic components. However, the use of liquid inside the channels makes the system bulky and difficult to handle.

The PCM emerges as a better cooling technique that absorbs the amount of heat dissipated by the IC chips during melting. The absorbed heat is released when the chips are not under working condition and kept idle. Depending on the safe operating temperature of the system, the selection of the PCMs is carried out based on their melting point. On the application front, the PCMs are widely used in varied engineering domains like photovoltaic systems, solar power plants, space industries, waste heat recovery systems, preservation of food and pharmaceutical products, building cooling, etc. However, the present study only focuses on the cooling of the IC chips using the air and PCM. Hence these two cooling methods are discussed extensively under Sections 1.4.1 and 1.4.2, respectively.

1.4.1 Air Cooling

Air continues to be the most widely used coolant in electronic systems due to its low cost, ease of availability, less maintenance, non-contamination, and, most importantly, it doesn't add vibration, noise, and humidity to the system. The method doesn't have any fluid-handling problems. This method is generally free from freezing, boiling, and dripping problems, and is generally preferred for low to medium heat flux levels (100–10,000 W/m²). This can be used for varied engineering applications like avionics, cooling of computers and data centres, automobile electronics, etc. Moreover, this takes into account the combined heat transfer modes (conduction, convection, and radiation). However, natural, forced, and mixed convection is the commonly used air cooling technique.

Most of the consumer-based electronic products and low-end applications of electronics deal with heat dissipation in the form of natural convection and surface radiation. However, natural convection has an edge over radiation due to its low operating cost, high system reliability, and noise-free operation. Here, the buoyancy forces due to the density difference cause the fluid movement. Natural convection cooling is characterized by non-dimensional parameters like Grashof number (Gr) and Rayleigh number (Ra). However, for higher cooling rates of the electronic components, natural convection air cooling is not preferred as a result of which, forced convection air cooling is achieved using external means like a fan or blower. Here the buoyancy forces are negligibly small. The forced convection cooling is characterized by Reynolds number (Re).

However, in many air-cooled systems, mixed convection is preferred and has been the area of interest for many researchers. Here both the external inertia forces and the buoyancy forces are of the same order. Mixed convection cooling is employed in different applications like cooling of heat exchangers, electronic components, avionic packages, datacentres, etc., and can be used for higher heat transfer rates up to 16,000 W/m². The mixed convection cooling is characterized by the non-dimensional parameter Richardson number (Ri). The Ri $\geq$ 1 leads to pure natural convection, Ri $\leq$ 1 indicates the forced convection, and Re $\equiv$ 1 identifies the mixed convection heat transfer regime.

1.4.2 Phase Change Material-Based Cooling

The PCMs are receiving enormous attention for the cooling of the electronics due to its high value of both heat capacity and latent heat of fusion leading to improved thermal energy storage. The PCM is in solid form at the initial stage, then starts melting, and becomes liquid with the increase in temperature of the electronic components. The PCM again changes its phase from liquid to solid after the removal of heat from the components and a drop in their temperature. This feature of phase interchangeability and reversibility without getting vaporized is an important characteristic of PCM. Hence, the use of PCM is gaining interest in the domain of electronic cooling, battery thermal management, building cooling, solar power generation system, etc.

Several PCMs are available in the market having a wide range of thermal conductivity, latent heat, heat capacity, and melting temperature. The classification of PCMs (shown in Figure 1.4) depends on their melting temperature and ease of application. The different types of PCM include paraffin wax, non-paraffin organics, hydrated salts, and metals. Paraffin wax is the preferred candidate for electronic applications due to its properties like high latent heat value, a wide range of melting points, non-corrosive, chemically inert, and negligible volumetric change during the phase change. Hydrated salts are generally used for large energy-storage applications, as they are much cheaper. However, the component design using the hydrated salts must take into account the effect of corrosion; hence, a limited number of cycles are available for the same. During the melting of the PCMs, the sensible heat causes the initial temperature rise of the PCM; the energy is then stored in the form of latent heat after reaching the PCM melting-point temperature. The total of the sensible heat and latent heat leads to the total energy stored in the PCM. Hence, the PCM is extensively used for electronic cooling applications, as it stores the thermal energy apart from cooling the system.

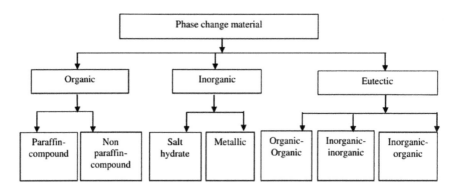

FIGURE 1.4 Different categories of phase change materials (PCMs). (Sharma et al., 2009.)

1.5 OPTIMIZATION IN HEAT TRANSFER

The optimization techniques in the heat transfer domain have received enormous attention in recent years due to the rapid progress in computer technology and the development of new and powerful optimization algorithms. Optimization is a process of determining the conditions that give the extremum (maximum or minimum) value of a function. The optimization methods are broadly classified as calculus and search-based methods. Each method again deals with either single or multi-variable problems, which may be either constrained or unconstrained in nature.

For problems like heat attenuation and liberation from the electronic components, the solutions are very sensitive to the variations in their component geometry, where convection is the dominant heat transfer mode. Hence, this affords tremendous scope in geometric optimization. Several techniques like the analytical approach, downhill simplex method, gradient descent, and many more are available for the optimization of the heat transfer systems. However, they have a general tendency to get "stuck" at a local minimum. To overcome this, evolutionary algorithms like genetic algorithms, simulated annealing are gaining popularity and are proposed in the present study.

CHAPTER 2

State-of-the-Art Studies in Electronic Cooling

2.1 INTRODUCTION

This chapter elucidates the complete review of the literature pertinent to the present study. For easy understanding, the literature review has the following subsections:

- Studies pertaining to natural convection cooling of discrete integrated circuit (IC) chips

- Studies pertaining to forced and mixed convection cooling of discrete IC chips

- Studies pertaining to phase change material (PCM) based cooling of discrete IC chips

2.2 STUDIES PERTAINING TO NATURAL CONVECTION COOLING OF DISCRETE IC CHIPS

The experimental and numerical analysis of protruding heat sources cooled using air under the natural convection heat transfer mode was performed by (Fujii et al., 1996). The heat sources were mounted vertically on the parallel plate mimicking the IC packages. They reported the Nusselt number and modified the Grashof number correlation to predict the heat source temperature for their different height and plate spacing.

DOI: 10.1201/9781003188506-2

The optimum heat source spacing has lowered down their temperature at their mid-height.

The numerical investigation on three protruding IC chips mounted on the vertical and horizontal substrate board orientations (in both in-line and staggered manner) under the natural convection heat transfer mode was carried out by (Liu et al., 1997). They reported that the temperature of the IC chips was maximum for the vertical board orientation as compared to the horizontal one. They reported a correlation for the non-dimensional maximum temperature of the IC chips in terms of their Rayleigh number.

Aydin and Yang (2000) conducted the 2D numerical simulations on four flush-mounted heat sources kept inside an enclosure. They varied the Rayleigh number in the ranges of 10^3 to 10^6 and concluded that the conduction was dominant at the lower Rayleigh number, and it has a negligible effect at the higher value. There was a significant temperature drop of the heat sources with the increase in their Rayleigh number.

Bessaih and Kadja (2000) performed numerical simulations to study the conjugate heat transfer of three ceramic heat sources mounted vertically under the natural convection heat transfer mode. They varied the distance between the heat sources and suggested that a better cooling rate was achieved by increasing the heat source distance.

The numerical simulations for three protruding IC chips arranged in five different positions inside an enclosure were conducted by (Chuang et al., 2003) under natural convection. The objective was to study the effect of temperature on heat sources. The maximum temperature was observed for the upper chip placed in the vertical arrangement as compared to the horizontal one.

Tou and Zhang (2003) performed numerical simulations to study the heat transfer characteristics of nine flush-mounted heat sources mounted vertically and kept inside a dielectric fluid-filled 3D enclosure under the natural convection heat transfer mode. The study has predicted the substrate board orientation (0° to 360°) effect on the heat transfer characteristics of the heat sources. They concluded that a 230°- to 310°-oriented enclosure had a negligible effect on the cooling of the heat sources.

Bhowmik and Tou (2005) performed the transient experiments on four flush-mounted heat sources placed vertically inside an enclosure using water as a coolant under the natural convection heat transfer mode. The effect of different heat fluxes and the heat source geometries on their temperature were studied. They proposed a Nusselt number equation in terms of the Fourier number and Rayleigh number and observed that the Nusselt number of the heat sources has decreased and the Rayleigh number has increased for their power ON condition.

The numerical simulations on four flush-mounted heat sources placed inside an enclosure under the natural convection heat transfer mode were carried out by (Bazylak et al., 2006). The heat source spacing and their length were varied to study the effect of different non-dimensional parameters on their heat transfer rate. They observed that increasing the spacing between the heat sources gave a maximum heat transfer rate.

Desrayaud et al. (2007) carried out the numerical simulations for a protruding heat source mounted between two parallel PCB stacks. They varied the thermal conductivity and the thickness of the PCB and concluded that the heat source temperature was decreased significantly by increasing the PCM thermal conductivity.

Florio and Harnoy (2007) performed the numerical study under the natural convection heat transfer mode on a protruding heat source with an oscillating plate placed above it. The spacing between the plate and the heat source was studied by varying the distance between these two. It was observed that the oscillating plate with an amplitude of 1/3rd spacing and a frequency of 2π has helped in enhancing the cooling of the heat source.

Bakkas et al. (2008) carried out the numerical simulations on a protruding heat source placed in the simple and double domain. The heat source was supplied with uniform heat flux by varying its height along with the Rayleigh number. They suggested that increasing the height of the heat sources has increased their Nusselt number and thus the Rayleigh number has increased. They concluded that the use of a double domain was more significant than a simple domain as the heat transfer characteristics were more evolved in the double domain. They have also proposed an equation for the Nusselt number in terms of their Reynolds number.

The numerical simulations on the flush and protrude-mounted heat sources placed in the triangular enclosure were carried out by (Koca et al., 2008). They varied the positions of the heat sources along with their height and the Rayleigh number. They observed the maximum heat transfer by placing the protruding heat sources near the inclined wall of the triangular enclosure. The Nusselt number (Nu) of the heat sources has increased with the increase in their Rayleigh number, and the interaction between the heat sources was found to be significant for their less spacing leading to enhance their temperature.

Narasimham (2010) reviewed the natural convection heat transfer as an inexpensive, highly dependable, vibration and noise-free mechanism for the heat removal from the electronic packages, and can accommodate the heat fluxes of the order of 10^3 W/m² without exceeding the package temperature beyond 80–90°C.

The experimental and numerical simulations on five heat sources arranged in 5 × 5 arrays and supplied with non-uniform heat fluxes were carried out by (Sudhakar et al., 2010b). They studied the heat transfer phenomenon for different heat source spacing and concluded that the heat source temperature has decreased with the increase bin their spacing. The spacing of the heat sources, their size, and their location on the substrate board were governed by a dimensionless distance parameter, λ.

Nardini and Paroncini (2012) performed the numerical and experimental analysis on four protruding heat sources mounted horizontally at various positions on the substrate board. The experiments were conducted in the square enclosure using holographic interferometry. They suggested that the conduction and convection effects were significant for Rayleigh number values of 10^4 and 10^5, respectively. The temperature of the heat sources was highly influenced by their size and location and found that the numerical results were in strong agreement with the experiments for the Rayleigh number range of $10^4 \leq Ra \leq 10^5$.

Mahdi (2013) conducted the experiments to investigate the influence of heating the lower half and cooling the upper half of the vertical surface of a rectangular enclosure under the natural convection heat transfer mode. They carried out 15 tests for three different volumetric flow rates and five different heat fluxes. They suggested that, with the increase in the volumetric flow rate of water, the temperature ratio of the lower and upper half of the surface has decreased.

Habib et al. (2014) conducted experiments for the uniform and non-uniform spacing of the flush-mounted heat sources by varying their heat input (10, 15, and 20 W) and Rayleigh number (9.06×10^9 to 1.41×10^{10}). They studied the effect of the Rayleigh number on the velocity and temperature of the heat sources inside the enclosure. They suggested that the non-uniform spacing of the heat sources has led to lower their temperature as compared to their uniform spacing, and thus led to their better rate of cooling.

Talukdar et al. (2019) performed the numerical simulations for the symmetric and staggered arrangement of the three heat sources mounted in a vertical channel under the natural convection heat transfer mode. They found that the cooling of the IC chips was enhanced by 7–10% for the staggered arrangement in comparison to the symmetric one. This was applied to both uniform and non-uniform arrangements of the heat sources.

Roy et al. (2020) performed the numerical simulations on multiple trapezoidal heat sources inside a square cylinder. They varied the length, width, and height of the heat sources to study the effect on the eddies formation, Nusselt number, and Rayleigh number. They found that by increasing the length and width of the heat sources; their Nusselt number has decreased at the surface, but the overall Nusselt number has improved.

For convenience, the earlier literature review pertinent to the IC chip's cooling under natural convection heat transfer is presented in a tabular form, as given in Table 2.1.

TABLE 2.1   Summary of Literature Pertinent to the Natural Convection

S. No.	Author and Year	Parameters	Key Points
1.	Fujii et al. (1996)	Multiple heat sources arranged parallel on a vertical plate	2D governing equations are solved numerically and validated with the experimental results. The geometric parameters and heat transfer characteristics were studied for the practical case of an IC package.
2.	Liu et al. (1997)	Three protrude IC chip Inline and staggered arrangement for vertical and horizontal orientation board	The finite element method was used to solve numerically the effect of conjugate heat transfer on different orientations of the board. Horizontal board result in less temperature of IC chips
3.	Bessaih and Kadja (2000)	Three ceramic protrude heat sources Spacing and power cut-off was varied	The finite volume method (FVM) was used to solve turbulent flow under natural convection
4.	Aydin and Yang (2000)	Four flush-mounted heat sources	2D numerical simulation conducted under natural convection Heat transfer increases with the increase in Rayleigh number
5.	Tou and Zhang (2003)	Nine flush-mounted heat sources Dielectric fluid filled in the enclosure	Developed a correlation for Nusselt number with the different inclination and Rayleigh's number Heat transfer is non-uniform in the nine discrete heat sources
6.	Chuang et al. (2003)	Three protrude chips Five positions vertical, horizontal, staggered, triangular, and inverse triangular	The temperature of the chip is maximum for vertical arrangement Better cooling is obtained for horizontal arrangement

(Continued)

TABLE 2.1 *(Continued)*   Summary of Literature Pertinent to the Natural Convection

S. No.	Author and Year	Parameters	Key Points
7.	Bhowmik and Tou (2005)	Four flush-mounted heat sources Water used as a coolant in the enclosure	Different heat flux was supplied – 1,000 W/m² to 6,000 W/m² Correlation for Nu in terms of Γ_δ and R_δ
8.	Bazylak et al. (2006)	Four flush-mounted heat sources Variation of spacing and length of heat sources	Heat transfer increased with the spacing and length of heat transfer Rayleigh number increased with the length of the heat source
9.	Florio and Harnoy (2007)	One protrude-mounted heat sources Oscillating plate	Better cooling was achieved by the oscillation of the plate placed above the heat source
10.	Desrayaud et al. (2007)	One protruded heat source Placed between two parallel PCB stacks	Studied numerically, using FVM, the heat transfer characteristics on the increase in thermal conductivity and thickness of the substrate board
11.	Bakkas et al. (2008)	Four protrude heat sources Use of single domain and double domain	Studied numerically 2D by varying the height and Rayleigh's number Proposed a correlation for single domain and double domain between Nu vs. Ra
12.	Koca et al. (2008)	Combination of flush and protrude heat sources Triangular enclosure	Variation in spacing, height, and Ra Heat transfer increased with the height and spacing of the heat sources
13.	Sudhakar et. al. (2010b)	Five protrude-mounted heat sources 5 × 5 array for placing heat sources	Temperature decreases with an increase in λ Nusselt increases with the Grashof number Numerical results validated with experiment results
14.	Nardini and Paroncini (2012)	Four protrude heat sources Different cases of the arrangement of heat sources	Holographic interferometry used for experimental visualizations Nusselt number increase with Rayleigh number Better cooling is achieved at a higher Rayleigh number
15.	Habib et al. (2014)	Flush mounted discrete heat source Non-uniform and uniform heat flux, 2D, symmetric and uniform spacing	The non-uniform spacing of the heat sources shows a better cooling rate in comparison with the uniform spacing of heat sources

(Continued)

TABLE 2.1 *(Continued)*   Summary of Literature Pertinent to the Natural Convection

S. No.	Author and Year	Parameters	Key Points
16.	Mahdi (2013)	Flush mounted discrete heat source, vertical arrangement, non-uniform heat flux, 2D, symmetric and uniform spacing	Nusselt number increases with the increase in Rayleigh number and the better cooling rate is obtained at a higher volumetric flow rate of water
17.	Narasimham (2010)	Natural convection Symmetric, flush, and protrusion mounted	No attempts have been made to compare the results of 2D and 3D numerical analyses of natural convection, with respect to medium and low aspect ratio enclosures
18.	Karvinkoppa and Hotta (2017; 2019)	Air cooling Phase change material	Conducted simulation on IC chips under natural convection and using PCM based mini-channels

2.3 STUDIES RELEVANT TO FORCED AND MIXED CONVECTION COOLING OF DISCRETE IC CHIPS

Choi and Ortega (1993) performed numerical simulations to study the effect of channel orientations for the cooling of the flush-mounted heat sources under the natural, mixed, and forced convection heat transfer mode. They rotated the channel from 0° to 360° and varied the non-dimensional parameters between $10^3 \leq Gr \leq 10^5$ and $0.1 \leq Re \leq 5000$. They observed that the Nusselt number variation was significant for different channel orientations under the natural and mixed convection heat transfer mode; however, the effects were negligible under the forced convection heat transfer mode.

Du et al. (1998) performed the 2D numerical simulations for three non-identical protruding heat sources placed vertically inside the channel. The numerical simulations were carried out under the mixed convection heat transfer mode using a SIMPLER algorithm with Boussinesq approximation. They found that the heat source spacing was vital for their cooling and suggested that the heat sources must be placed at the channel entrance for their better cooling.

Queipo and Gil (1999) performed a numerical study on the multi-objective placement of the electronic components (heat sources)

considering their position and the PCB design. They combined the heat transfer study with the Pareto optimization to determine the optimal position of 36 electronic components to reduce their failure rate. The objective function of the optimization was formulated using the Arrhenius relation. They obtained the best arrangement for placing the electronic components using the Pareto optimal solution for which the component failure rate has reduced.

Ozsunar et al. (2001) performed the numerical simulations to study the effect of channel inclination, Grashof number, and Reynolds number on the mixed convection heat transfer mode inside a rectangular channel. They supplied uniform heat flux at the bottom of the channel with its inlet and outlet walls open and its top wall was adiabatic. The heat transfer has enhanced with the channel inclination which ultimately increased the Nusselt number and Grashof number.

Chen and Liu (2002) conducted the experiments on nine protruding heat sources mounted in a wind tunnel under the forced and mixed convection heat transfer mode. They reported that the uniform heat source spacing was less effective as compared to their centre-to-centre placement. They found that, for a spacing ratio of 1.8, the maximum heat source temperature was found to be 8.24% more than the uniform spacing.

Da Silva et al. (2004) performed the analytical and numerical simulations for the flush-mounted heat sources of different sizes under the forced convection heat transfer mode. The objective was to increase the overall conductance of the heat sources at different Reynolds numbers and by placing these at different positions. They observed that the placement of the heat sources was dependent on the Reynold number and the overall conductance value has increased by placing the heat sources coarsely. The temperature of the heat sources was minimum by placing the heat sources at the channel entrance.

Bhowmik et al. (2005) conducted the experiments on four flush-mounted heat sources placed inside an enclosure using water as the interacting fluid. The flow rate, heat flux, and the geometric spacing of the heat sources were varied to study the effect of the heat transfer characteristics of the heat sources under various heat transfer modes. A correlation was developed for the Nu in terms of their Gr and Re.

Premachandran and Balaji (2005) studied numerically the cooling of four protruding heat sources under the mixed convection heat transfer mode using water and FC 77. The spacing, height, and size of the heat sources were fixed and the effect of thermal conductivity ratio between the heat sources (k_p) and the fluid (k_f), between the heat sources and the substrate board (k_s), Reynolds number, and Grashof number on their heat transfer were studied. They found that thermal conductivity has played a key role to predict the heat source temperature, and the maximum temperature was increased by 40% for the thermal conductivity ratio of $1.032 \leq k_p/k_f \leq 10.32$.

Guimarães and Menon (2008) carried out the numerical simulations on 3 flush-mounted heat sources placed inside a channel under the combined free and forced convection heat transfer mode. They varied the Reynolds number, Grashof number, and the channel inclination to study the effect of these parameters on the temperature of the heat sources. They found that the Nusselt number of the heat sources has played a significant role in enhancing their heat transfer rate for the channel inclination of 45° and 90° and at a lower Reynolds number.

Madadi and Balaji (2008) performed the optimization studies for the three flush-mounted heat sources placed randomly inside the square enclosure and supplied with uniform heat fluxes under the forced convection heat transfer mode. They obtained the temperature of the heat sources by carrying out the numerical simulations using Fluent to minimize their maximum temperature. They used the numerical data-driven combined artificial neural network (ANN) and genetic algorithm (GA) technique to obtain the optimal heat source position.

Aminossadati and Ghasemi (2009) carried out the numerical simulations on a single flush-mounted heat source placed inside a cavity (at its left, bottom, and right walls). They varied the cavity aspect ratio, location of the heat source, and the Richardson number to predict the temperature of the heat sources. They concluded that the heat transfer rate of the heat sources was significant for the aspect ratio of two, where the Richardson number was higher due to the buoyancy-induced flow.

Ozsunar et al. (2009) performed the numerical simulations using ANSYS Fluent for a single chip under the different input parameters like power input, chip material (copper, aluminium, and silicon), and their thickness. The numerical results were used to power the ANN that has reduced the simulation time. They found that the temperature of the chip was depended on all the earlier mentioned parameters, and the ANN model was used to predict the chip temperature accurately.

Amirouche and Bessaih (2012) carried out the numerical simulations for 10 heat sources mounted inside a vertical channel under the mixed convection heat transfer mode. They varied the non-dimensional parameters (Gr, Re, and Ri), and the heat input to study the effect of these parameters on the temperature of the heat sources. Different combinations of the power inputs (ON and OFF conditions) to the heat sources were considered for the analysis. They suggested that when few heat sources were kept OFF during the simulation, due to the conjugate heat transfer, their temperature was increased, thereby decreasing the heat source temperature under the ON condition.

Panthalookaran (2010) performed the numerical simulations for an enclosure consisting of the transformer, array of metal-oxide-semiconductor field-effect transistors (MOSFETs), heat sinks with vertical and horizontal slots under the different heat flux values. The objective was to improve the cooling rate of the entire system using the various components, and the slots for the exit of air. They found that the vertical arrangement of the slots gave a better heat dissipation rate than the horizontal one. The temperature of the MOSFETs and other components were brought down using the heat sinks. They concluded that the outlet vent of the enclosure must be greater than or equal to the inlet vent.

Sudhakar et al. (2010a) performed the experimental and numerical analysis on 15 protruding heat sources arranged in a 5 × 3 array under the mixed convection heat transfer mode. They employed the numerical data-driven combined ANN-GA-based technique to obtain the optimal position and the temperature of the heat sources. They observed that the temperature of the heat sources has increased from the upward to the centre position due to more interaction, and the heat sources must be supplied with optimal heat flux to keep their temperature below the critical value.

Kargar et al. (2011) performed the 2D simulations for two flush-mounted heat sources kept inside an enclosure using the copper-based water nanofluid as a coolant. They varied the position of the heat sources and the nanoparticle volume fraction and observed that, with an increase in the Rayleigh number and the volume fraction of the nanoparticles, there was a significant temperature drop of the heat sources. They suggested that the optimal volume fraction and position of the heat sources have led to a better rate of cooling.

Ajmera and Mathur (2015) conducted the experiments on three flush-mounted heat sources for multi-ventilated enclosure under the mixed convection heat transfer mode. The three heat sources are placed beside each other to study the effect of various non-dimensional parameters like Re, Ri, and Gr. The non-uniform heat flux was supplied to all the heat sources, and it was observed that the heat sources near the enclosure inlet were cooled fast as compared to the other two heat sources for all the Reynolds numbers. The heat sources with higher heat flux must be placed at the enclosure inlet, and the lower heat flux must be placed at the enclosure outlet.

Ahmed et al. (2016) performed the numerical simulations on three protruding heat sources kept inside an inclined channel under the mixed convection heat transfer mode. They varied the channel angle between 0° and 90° to study its effect on the heat source temperature for various Reynolds number. The better cooling rate was obtained at 45° and 90°. The effect of the channel inclination was significant for the higher velocities that have a better heat transfer rate of the heat sources.

Durgam et al. (2018) carried out the experimental and numerical investigations on seven protruding heat sources using a dummy heat source under the forced convection heat transfer mode. The effect of the heat source of temperature on the substrate thermal conductivity was studied using different air velocities. They proposed a correlation for the heat source temperature in terms of the substrate thermal conductivity, Reynolds number, and the spacing between them.

Chaurasia et al. (2019) performed the experimental and numerical analysis for six protruding heat sources by varying different parameters like velocity, substrate thermal conductivity, emissivity, and heat source positions both in the stream-wise and span-wise directions. They found that the heat transfer due to radiation plays a significant role at higher temperatures and its influence was neglected when there was an increase in the thermal conductivity of the substrate board and Reynolds number.

Durgam et al. (2019) conducted experimental and numerical investigations on 15 protruding heat sources under the laminar forced convection heat transfer mode by supplying three different uniform heat fluxes with different velocities. They used three different substrate materials, i.e. FR-4, bakelite, and copper cladding, to study the effect of temperature on the thermal conductivity of the substrate board. They have developed a correlation on these parameters to study the temperature variation of the heat sources. They suggested that better cooling of the heat sources was achieved with an increase in the thermal conductivity of the substrate board and Reynolds number. The heat sources generating more heat must be placed at the substrate bottom.

Garcia et al. (2019) carried out the numerical simulations on the inclined channel with two protruding heat sources placed parallel to each other on the top and bottom walls of the channel, respectively. The transient simulations were carried out to study the influence of buoyancy and Reynolds number on the different channel inclinations. They suggested that the buoyancy effect becomes negligible for the horizontal orientation and significant for the vertical channel orientation. The heat transfer rate has increased with the increase in Reynolds number and for the fixed Richardson number value.

Korichi and Laouche (2019) performed the numerical analysis on the pulsed flow inside a vertical channel on the four protruding heat sources under the mixed convection heat transfer mode. The flow inside the channel was varied with the change in amplitude. They suggested that the heat transfer characteristics and the flow behaviour were strongly affected by the variation in air amplitude. The Nusselt number has increased drastically for the higher amplitude or pulsation which was not appropriate for the steady-state results. They concluded that the proper air amplitude was required for the cooling of the electronic components.

Kumar and Phaneendra (2019) performed the numerical simulations on ten protruding heat sources placed inside the duct under the mixed convection heat transfer mode. The ten heat sources were placed using a coordinate system and 40 configurations were considered for the analysis by supplying a uniform heat flux of 25×10^4 W/m². They found that, by placing the heat sources close to each other, their temperature has increased due to the increased thermal interaction between them. They concluded that the heat sources must be far from each other, and the heat source generating more heat must be placed at the duct inlet.

For convenience, the earlier literature review pertinent to the cooling of the IC chips under the forced and mixed convection heat transfer mode is presented in a tabular form, as given in Table 2.2.

TABLE 2.2 Summary of Literature Pertinent to the Forced and Mixed Convection

S. No.	Author and Year	Parameters	Key Points
1.	Choi and Ortega (1993)	Natural, mixed, and forced convection, Single flush-mounted heat source (FMHS)	Different orientation of channel 0 to 360° $10^3 \leq Gr \leq 10^5$ and $0.1 \leq Re \leq 5,000$ Mixed convection gave a better rate of cooling
2.	Du et al. (1998)	Mixed convection, Three protrude-mounted heat sources (PMHS) mounted vertically, Identical HS	Studied numerically using a SIMPLER algorithm $0 \leq Ra \leq 10^7, 0 \leq Re \leq 200$ The higher power rating heat source must be placed at the entrance of the channel for better cooling
3.	Ozsunar et al. (2001)	Mixed convection heat transfer, uniform heat flux supplied to the bottom, Identical HS	The bottom wall of the channel was heated to study the effect on mixed convection heat transfer for different Gr, Re, and θ. $7 \times 10^5 \leq Gr \leq 4 \times 10^6, 50 \leq Re \leq 1000, 0° \leq \theta \leq 90°$
4.	Chen and Liu (2002)	Forced and mixed convection, 3×3 PMHS, Identical HS	Spacing from the centre to centre distance has maximum heat transfer from the heat sources Conventional uniform spacing does not reduce the temperature The temperature difference of IC chips was reduced by 27%
5.	Guimarães and Menon (2003)	Mixed convection, Single FMHS	The substrate board is oriented at different angles. Non-dimensional parameter ranges between $1 \leq Re \leq 500, 10^3 \leq Gr \leq 10^5$, and $0° \leq \gamma \leq 90°$
6.	Bhowmik et al. (2005)	Natural, mixed, and forced convection, 4 FMHS, Water coolant	Experiment study was carried out for a different range of $40 \leq Re \leq 2,220$ based on heat source lengths ranging from 50 to 2,775 mm, respectively. The heat flux ranges from 0.1 W/cm² to 0.6 W/cm², and a hydraulic diameter ranging from 40 to 2,220
7.	Premachandran and Balaji (2005)	Mixed convection Four PMHS, Identical HS Water and FC70 used as the coolant	Numerically investigated using FVM to study the effect of thermal conductivity ratio, spacing ratio, Re, and Gr on the heat transfer characteristics The buoyancy effects were significant in water and negligible for FC 70
8.	Guimarães and Menon (2008)	Three FMHS, Mixed convection, Identical HS	Studied numerically the effect of $1 \leq Re \leq 1,000, 10^3 \leq Gr \leq 10^5$, and $0° \leq \gamma \leq 90°$ on temperature A better rate of heat transfer was achieved at 45°

(Continued)

TABLE 2.2 *(Continued)*   Summary of Literature Pertinent to the Forced and Mixed Convection

S. No.	Author and Year	Parameters	Key Points
9.	Aminossadati and Ghasemi (2009)	Mixed convection, Single FMHS	Numerically studied the effect of placing the heat source at different locations and different aspect ratios. The higher heat transfer observed at cavity aspect ratio 2, where Ri is more significant for buoyancy induced flow. The Ri and aspect ratio varied between 0.1 and 100 and 1 and 5, respectively
10.	Amirouche and Bessaih (2012)	Ten PHMS, Mixed Convection, Identical HS	Conducted numerical simulation on ten PMHS with few PMHS no-power conditions and remaining on $Gr = 6.40 \times 10^5$ and 12.80×10^5 $Q = 0.2$ W and 0.4 W, $Re = 500, 750, 1,000$, $Ri = 0.64, 1.14, 2.56$
11.	Satish Kumar Ajmera (2015)	3 FMHS, Mixed convection, Identical HS	Experimental investigation for aspect ratio AR (L/H) = 1, range of Reynolds number and Grashof number considered are $270 \leq Re \leq 6274$ and $7.2 \times 10^6 \leq Gr \leq 5.5 \times 10^7$ and $0.201 \leq Ri \leq 571$ Higher heat flux heat source should be placed at the inlet
12.	Ahmed et al. (2016)	Mixed convection, Three PMHS, Identical HS	The inclination is dominant for higher Re Heat transfer is more at 45° for $Ri = 1$ $1 \leq Re \leq 100, 10^3 \leq Gr \leq 10^6, 0.1 \leq Ri \leq 10$, and $0° \leq \gamma \leq 90°$
13.	Durgam et al. (2018)	Forced convection, 7 PMHS and 8 dummy heat sources, Non-identical HS	An experimental investigation was conducted T_{max} decreases with an increase in λ Substrate copper cladding gives better cooling of heat sources $1.3 \leq \lambda \leq 2.38, 11.21 \leq k \leq 330$, and $1125 \leq Re \leq 1,825$
14.	Madadi and Balaji (2008)	3 FMHS, Forced convection, Identical HS	Numerically investigated for velocity 4 m/s Used ANN + GA to determine the optimal position of the heat sources
15.	Kumar and Phaneendra (2019)	10 PMHS, Mixed convection, Identical HS	Numerically investigated for the heat flux of 25×10^4 W/m² and suggested that a higher temperature heat source must be placed at the inlet
16.	Sudhakar et al. (2010b)	Mixed convection, 15 FMHS, Identical HS	Optimal heat flux must be supplied for operation of the heat source below the critical temperature

(Continued)

TABLE 2.2 *(Continued)*   Summary of Literature Pertinent to the Forced and Mixed Convection

S. No.	Author and Year	Parameters	Key Points
17.	García et al. (2019)	Mixed convection, Two PMHS, Identical HS	Transient numerical studies were conducted for different parameters $0° \leq \gamma \leq 90°$, $100 \leq Re \leq 1,000$ Studied the influence of buoyancy on the different inclinations of the channel Strouhal and Nusselt were calculated for different Re and Ri
18.	Korichi and Laouche (2019)	Mixed convection, Four PMHS, Identical HS	Numerical study on pulsation in the flow under mixed convection The temperature and flow behaviour are affected due to pulsation in the flow
19.	Durgam et al. (2019)	Forced convection 3×5 PMHS Identical HS	Experiment and numerical study for 15 PMHS $1,000 \text{ W/m}^2 \leq q \leq 3,000 \text{ W/m}^2$, $0.6 \text{ m/s} \leq v \leq 1.4 \text{ m/s}$ The temperature of HS drops with an increase in thermal conductivity of the substrate board
20.	Chaurasia et al. (2019)	Mixed convection, 6 PMHS	Numerical and experimental study The ranges of Reynolds number, emissivity, and thermal conductivity of PCB are 115–690 (corresponding inlet velocity of 0.25–1.5), 0–0.9, and 0.038–1.4 W/mK, respectively The temperature of HS drops with an increase in spacing stream-wise and span-wise, thermal conductivity, and Re
21.	Queipo and Gil (1999)	Multi-objective optimization 36 PMHS	The numerical study was conducted with Pareto optimization The thermal and non-thermal objective functions of the optimization were formulated using the Arrhenius relation
22.	Panthalookaran (2010)	Natural convection transformer, heat sink, an array of MOSFET, vents at inlet and outlets	A numerical study was conducted with MOSFETs-supplied power input of 10 W Better cooling for the vertical slot of the array than horizontal The use of a heat sink helps in the cooling of components
23.	Kargar et al. (2011)	Optimization, CFD + ANN, two FMHS copper-water nanofluid	Numerical investigation of the effect of temperature on the volume fraction of nanoparticles and position of heat sources Rayleigh number varied between 10^3 and 10^6 and volume fraction of fluid between 0 and 0.1 CFD-assisted ANN to increase the simulation speed and obtain accurate and precise results

(Continued)

TABLE 2.2 *(Continued)*   Summary of Literature Pertinent to the Forced and Mixed Convection

S. No.	Author and Year	Parameters	Key Points
24.	da Silva et al. (2004)	Forced convection FMHS	Analytical and numerical study Temperature reduces with an increase in Re Overall conductance increased by 25% for $10^2 \leq Re \leq 10^4$
25.	Ozsunar et al. (2009)	Single-chip PMHS CFD + ANN	A numerical study was conducted for different $2 \leq Q \leq 15$ The temperature was affected significantly by the material of chip, thickness of chip, and power input supplied to the chip

2.4 STUDIES PERTAINING TO THE PHASE CHANGE MATERIAL-BASED COOLING OF DISCRETE INTEGRATED CIRCUIT CHIPS

Zhou et al. (2001) performed the experiments on three protruding heat sources placed inside the n-octadecane (melting point, Tm: 28°C)-filled rectangular enclosure. They have analyzed the effect of melting, sub-cooling, and the Stefan number, and observed that, at the initial PCM-melting stage, the effect of sub-cooling has appeared, and then it has disappeared when its melt fraction has reached 0.5.

Fatih Demirbas (2006) carried out an extensive review of various PCMs used for the different electronic cooling applications. The melting temperature of different PCMs was elaborated for the solar application. They concluded that the paraffins having low thermal conductivity were used for the solar cooling application and focused on their thermal energy storage.

Saha et al. (2006) and Saha and Dutta (2010) conducted experimental and numerical investigations on the PCM-based heat sinks to study their cooling effect on the volume variation of the thermal conductivity enhancer (TCE). The heat sinks were used as the TCE and n-eicosane and paraffin wax were used as the PCM. They varied the volume content of the TCE and the heat input to the PCM and found that an 8% volumetric fraction of the TCE with the maximum number of fins gave a better thermal performance.

Kandasamy et al. (2008) conducted an experimental and numerical study on a plastic quad flat package (QFP) chip with a paraffin wax (Tm: 55°C)-based heat sink placed on the top of it. They carried out the transient analysis to enhance the working cycle of the QFP by supplying different power inputs. They observed that the PCM has melted faster for higher power input (6 W), and better cooling performance was obtained for Q > 3 W. They performed the numerical analysis using the enthalpy porosity technique and found that the numerical results were in strong agreement with the experiments.

Faraji and El Qarnia (2010) carried out the 2D numerical simulations on three protruding heat sources kept inside an n-eicosane (melting point, Tm: 36°C)-filled enclosure. The objective was to study the PCM melt fraction and validate the numerical results with the literature. They supplied constant heat flux to all the heat sources and observed that the heat sources placed at the enclosure bottom had the lowest temperature, at the top had the maximum temperature followed by the centre, due to the combined effect of conduction and convection.

Zeng et al. (2010) performed an experimental analysis on silver nanowire composite PCM to enhance the thermal conductivity of the PCM (1-tetradecanol) without reducing its enthalpy content. They observed that the thermal conductivity of the composite PCM with nanowire has increased to 1.45 W/mK, and their enthalpy content has increased to 76.5 J/g. Differential scanning calorimeter (DSC) and the hot disc thermal analyzer were used to determine the enthalpy content and the thermal conductivity of the PCM, respectively.

Cui et al. (2011) performed the experimental analysis on the composite PCM using the carbon nanofibers and carbon nanotubes with soy wax and paraffin wax. The percentage volume content of the nanotubes and nanofibers was varied in both the PCMs to study the effect of percentage variation on the thermal conductivity of the composite PCM. Their objective was to increase the thermal conductivity of the composite PCM. They observed that the thermal conductivity of the carbon nanofiber with soy wax was increased from 0.324 to 0.469 W/mK and for the carbon nanotube, it has increased from 0.343 to 0.403 W/mK. They have observed similar trends for the paraffin wax as well.

Yavari et al. (2011) conducted the experimental analysis to enhance the thermal conductivity of the composite PCM using 1-octadecanol and nanostructured graphene. The percentage volume content of graphene was varied between 0 and 4%. They concluded that the composite PCM thermal conductivity has increased by 140%, and their heat capacity was reduced by 15.4% using 4% graphene.

Ye et al. (2011) conducted a numerical study using the paraffin wax (Tm: 33°C to 35°C) for the plate-fin application to study their thermal energy storage. They have developed an empirical correlation to predict the PCM temperature with respect to time. They found that the paraffin wax has stored the maximum thermal energy when their temperature difference was 20°C.

Baby and Balaji (2012) performed the experimental investigation on the PCM-based heat sinks using two different PCMs, paraffin wax (Tm: 53°C to 57°C) and n-eicosane (Tm: 36.5°C). They have analyzed the heat sink with the different pin-fin arrangements, power input levels of 5–8 W, volume fractions, and TCE. They found the enhancement ratio for the n-eicosane at 7 W with a volume fraction of 1. They obtained the optimal configuration (optimal volume fraction was 0.97 and the corresponding TCE was 1.097) of the pin-fin using the experimental data-driven combined ANN-GA technique.

Salunkhe and Shembekar (2012) carried out an exhaustive review of various PCMs and their characteristic behaviour on the thermal performance of the system. They have reported the challenges faced in implementing the micro-PCM encapsulation and the melting and solidification characteristics of different PCMs. The various dimensionless parameters associated with the solidification and melting of PCM were analyzed. They suggested that conduction and natural convection were significant for PCM melting.

Mahmoud et al. (2013) performed an experimental study on the PCM-based heat sinks using six different types of PCM under different power inputs. They compared the results using single, double, triple, and multiple cavities filled with PCM and found that the inclusion of PCM has a significant temperature drop of the heat sink. The PCM-based heat sinks were able to maintain the temperature below the critical value, and the PCM with a lower melting point has shown a longer operating cycle.

El Qarnia et al. (2013) performed the 2D numerical analysis using three protruding heat sources placed inside the n-eicosane-filled enclosure. The heat sources were supplied with different heat fluxes and their heat dissipation rate has helped in the PCM melting kept inside the enclosure. The heat dissipated from the heat sources was stored in the PCM during its melting that has led to a change in their melt fraction. The heat sources placed at the bottom have the lowest temperature as compared to the other two heat sources placed at the centre and the top, respectively.

Elmozughi et al. (2014) carried out the 2D and 3D numerical simulations using the sodium nitrate-stainless steel encapsulation PCM with

20% void air. The simulation was focused on the solar power system where the heat from the fluid was transferred to the PCM leading to the thermal storage in the PCM. They have carried out the simulations using ANSYS with the enthalpy-porosity technique and observed three significant zones – solid, liquid, and mushy. The effect of thermal and volume expansion of the PCM was studied on energy storage.

Ling et al. (2014) conducted the numerical and experimental analysis on the expanded graphite (EG) PCM used for the battery thermal management to keep their temperature below the safe limit. The EG composite paraffin wax and the other composite PCMs were used to study the effect of density and thermal conductivity on the PCM melting temperature. They suggested that a higher PCM melting temperature was not desirable, and a melting temperature between 40 and 45°C was effective. They reported that the battery thermal performance was improved by increasing their density.

Rathod and Banerjee (2014) conducted the experimental analysis to study the latent heat storage of paraffin wax in the shell and tube heat exchanger. They studied the melt fraction of the PCM concerning the energy stored in the PCM by varying the mass flow rate and the fluid inlet temperature. They concluded that the melting fraction of the PCM was increased with the decrease in mass flow rate and the increase in the fluid inlet temperature. They found that the conduction was more significant during the discharging of PCM and the PCM solidification time was more as compared to its melting.

Janarthanan and Sagadevan (2015) carried out a detailed review of the application of different PCMs on domains like solar power systems, solar energy storage, waste heat recovery system, thermal comfort system, building, and electronic cooling systems. The different types of PCMs were used for thermal energy storage. The use of PCMs has enhanced the performance of the system substantially due to their properties like chemically stable, non-flammable, environment-friendly, and reuse/recycle capability.

Hasan et al. (2016) performed the numerical and experimental investigations on PCM-based heat sinks and compared the results without PCM-based heat sinks. Three different PCMs were used and their performance was compared with each other. They found that the salt hydrate and paraffin wax have shown better thermal performance as compared to the milk fat. The working cycle of the heat sinks has enhanced for both the natural and forced convection heat transfer mode using the PCM. They found that there was an increase in the working time of milk fat as compared to the other PCMs.

Ashraf et al. (2017), Arshad et al. (2018), and Ali (2018) conducted an experimental analysis on the PCM based heat sinks to study the parametric investigation of six PCMs, their heat input values, number of fins, fin geometry, and fin arrangements. They found that for the heat sink without PCM, the square pin-fin with staggered arrangement gave better results. The highest enhancement ratio was obtained for SP-31 followed by n-eicosane, RT-44, and paraffin wax. Paraffin wax was suitable for the highest power input of 8 W. They varied the volume content of the PCM between 50% and 100%, and concluded that the heat sink with 100% volume content has shown the maximum enhancement ratio for the heat sinks.

Abokersh et al. (2018) reviewed different PCMs used for domestic solar systems. They reported the various heat enhancement techniques for the extended surfaces, heat pipes, high conductive additives using multiple PCMs, and recommended that the PCM with a melting temperature of 50–60°C was appropriate for the solar energy system to store the energy during its charging and can be utilized during its discharging.

Gharbi et al. (2018) carried out numerical simulations on PCM-based copper heat sinks. Two different configurations of the heat sink were considered, one is the conventional copper heat sink which was kept inside an enclosure filled with PCM, and the other is the copper enclosure which is heated with a single protruding heat source. They observed that the longer fins gave a better cooling performance and a simple copper enclosure with PCM was better for the lower power input of the heat sources.

Hasan and Tbena (2018a) carried out numerical simulations using three different PCMs in the microchannel heat sinks. Seven different cases were considered in placing the three different PCMs in each channel of the heat sink. They found that the PCM has led to a significant heat sink temperature drop as compared to the without PCM-based heat sink. The different combinations of placing the PCM in the micro-channels have been determined with respect to the PCM melting rate. The PCM present near the heat sink base has melted due to conduction, and the melt rate was strongly dependent on the PCM type. The lower melting temperature of the PCM placed near the heat sink base gave a better thermal performance in reducing their temperature.

Hasan and Tbena (2018b) carried out numerical simulations on the PCM-based heat sinks for the electronic cooling application. Three different geometries of the micro pin-fin heat sinks were used (square, triangular, and circular) with and without the use of PCM. They used n-octadecane and RT-44 for the analysis, and found that n-octadecane has shown maximum temperature drop for the unfinned heat sink and the circular finned heat sink gave the better thermal performance.

Kahwaji et al. (2018) conducted an experimental study for six different PCMs with three different blends of alkanes and commercially available paraffin wax. The thermophysical properties of all the PCMs were determined using the DSC and the melting range of each PCM was reported signifying the application in solar thermal management of electronic domains. The specific heat and latent heat of fusion were found to be higher for the alkanes as compared to the commercially available paraffin wax.

Loganathan and Mani (2018) developed a multi-criteria optimization model for the selection of the PCM depending upon a suitable application. Ten different PCMs were selected with a melting point ranging from 31°C to 80°C. The algorithm was developed using fuzzy logic and it was paired to other algorithms to obtain the optimized results. The optimum results have suggested that the RT-80 (Tm: 80°C) was better for the electronic cooling.

Rehman and Ali (2018) conducted the experimental analysis on paraffin wax-based heat sinks with copper foam. The porosity of the copper foam, PCM volume content, and their power input were varied to study the effect on the base temperature of the heat sink. There was a 9% reduction in the PCM-based heat sink maximum temperature using the copper porosity of 0.95 as compared to 0.97 at 8 W for a PCM volume fraction of 0.8. They have also carried out the charging and discharging of PCM and found that the discharging phase does not affect the PCM performance.

Usman et al. (2018) conducted the experimental analysis on the triangular pin-fin PCM-based heat sinks with different pin-fin arrangements. The power input to the heat sinks was varied between 5 and 8 W. The comparison was made with the unfinned heat sinks, finned heat sinks, and PCM-based pin-finned heat sinks. Three different PCMs – RT-35, RT-44, and paraffin wax – were used with a 90% volume fraction. The operating time of the heat sinks was enhanced using all the PCMs and RT-44 was found to be more dominant than other PCMs due to its highest latent heat storage value. The in-line arrangement of the pin-fin was found to be more effective for all the cases of PCMs.

Bondareva et al. (2019) carried out the numerical simulations on the n-octadecane-based heat sink with a single protruding heat source placed at the bottom of the heat sink. The effect of nanoparticles (Al_2O_3) concentration and the heat sink inclination was studied on the PCM melting.

They concluded that, with the increase in the nanoparticle concentration, the PCM melting has enhanced and the heat sink must be placed for the smaller inclination, leading to significant convective heat transfer and, ultimately, significant PCM melting.

Gasia et al. (2019) conducted the experiments on the heat exchanger using fins, wool, and the n-octadecane. The objective was to enhance the thermal energy storage by the heat dissipation from the heat exchanger. They observed that the use of the earlier materials and PCM has led to the heat enhancement from the heat exchanger and an increase in the thermal energy storage by 14%.

Hamza and Mustapha (2019) carried out the 2D numerical simulations on the n-eicosane (Tm: 36.5°C) with copper oxide nanoparticles kept inside a rectangular enclosure for the cooling of a single protruding microprocessor chip. The concentration of the nanoparticles was varied in the PCM and the melting rate of the PCM was accelerated through which the temperature of the microprocessor chip was brought down. The Nusselt number was increased tremendously with the use of nanoparticle PCM as compared to the regular PCM.

Kalbasi et al. (2019) carried out the numerical simulations on RT-27 (Tm: 30°C)-based heat sink by supplying 5000 W/m² and 1000 W/m². The objective was to determine the optimum number of fins, fin thickness, and the PCM volume required to maximize the working cycle of the heat source to reach the critical temperature of 87°C. They found that the optimal number of fins has decreased with the increase in fin thickness and height ratio of the fins. The decreased number of fins gave rise to the increased spacing to accommodate a large amount of PCM. Further, the heat source temperature has decreased and their working cycle has enhanced. They have developed a correlation between the optimum number of fins, PCM volume fraction, and fin thickness.

Karami and Kamkari (2019) carried out the numerical simulations on the PCM-based pin-fin enclosure which was heated from the bottom for different inclinations. They observed that the melting time has reduced with the reduction in the inclination angle and adding fins to each inclination. The minimum melting time was obtained at 0°, and the melting time was also less for the vertical orientation as compared to other orientations.

Khademi et al. (2019) carried out numerical simulations on the oleic acid with water kept inside the enclosure to study the thermal energy storage. They carried out the numerical simulations under the extreme cold condition where the PCM was in solid form at –10°C for the different cases, 100% PCM in the enclosure, 50% PCM in the enclosure, and 50% water with 50% PCM in the enclosure. They observed that the mixture of PCM and water has accelerated the melting of PCM and helped in energy storage in extremely cold conditions. PCM was the primary fluid that does not dissolve in water and had a maximum latent heat of storage.

Rabie et al. (2019) conducted an experimental analysis on the photovoltaic cell using the PCM (RT-35HC)-based heat sink to increase the temperature distribution in the solar cell. The angle of inclination and the height ratio of the heat sink was varied to observe the effect of temperature on the PCM melting. They concluded that the inclination of the photovoltaic cell at 45° has shown less cooling performance due to the liquid PCM formation at the top layer of the heat sink. The maximum performance was obtained for a higher height ratio of the heat sink at 45°C.

Song et al. (2019) performed the numerical and experimental analysis on the cooling of the battery module system. The system consists of 106 cylindrical batteries supplied with different heat input which was placed on the cold plate. The cold plate acts as a heat spreader and has water flowing inside the mini-channels of the cold plate. The gap between the batteries was filled with paraffin wax. The temperature of the batteries was controlled by absorbing the heat by the PCM which helps in increasing the operating temperature and life cycle of the batteries.

Xie et al. (2019) carried out the numerical simulations on paraffin wax (Tm: 34 – 36.5°C)-based topology optimized tree-structured heat sink. The single protruding heat source was placed at the bottom and top of the heat sink. They found that the top-mounted heat sink and inverted heat sink have shown better performance, and the conduction heat transfer mode was dominant in the initial and secondary stages of PCM melting, whereas the convection mode of heat transfer was significant during the melting stage of the PCM due to diffusion.

Zarma et al. (2019) carried out the numerical simulations of the photovoltaic cell-based PCM (pure salt hydrate) with nanoparticles (Al_2O_3, CuO, and SiO_2) inside the heat sink. The objective was to study the effect of the percentage of concentration of the PCM nanoparticles to reduce the temperature of the solar cell. They reported that Al_2O_3 has shown better performance (5% concentration) and enhanced the PCM melting as compared to PCM in pure form.

For convenience, the earlier literature review pertinent to the PCM-based cooling of the IC chips is presented in a tabular form, as given in Table 2.3.

TABLE 2.3   Summary of Literature Pertinent to the Phase Change Materials

S. No.	Author and Year	Mode of Heat Transfer and Heat Sources	Key Points
1.	Solomon (1981)	Natural convection n-octadecane (Tm – 95°C)	Defined Stefan number $St = $ Sensible heat / Latent heat
2.	Fatih Demirbas (2006)	Different phase change materials (PCMs) Paraffin wax and hydrated salts are widely used	Reviewed different PCMs for various applications PCM is mainly used as a battery for thermal energy storage PCMs for the solar system used were Glauber's salt, calcium chloride hexahydrate, and paraffin wax
3.	Kandasamy et al. (2008)	Single PMHS QFP chip Paraffin wax (T_m – 55°C)	Experimental and numerical study PCM-based heat sinks supplied with power input $2 \leq Q \leq 6$ W They observed that PCM enhanced the cooling performance of QFP chips
4.	Faraji and El Qarnia (2010)	Three PMHS n-eicosane (T_m – 36°C) Vertical enclosure	2D numerical simulation under natural convection Melting of PCM is due to heat dissipation from the heat sources Bottom heat sources have the lowest temperature
5.	Ye et al. (2011)	Paraffin wax Plate fin system Paraffin wax (T_m – 33°C – 35°C)	3D numerical simulations Maximum energy storage at $\Delta T = 20$°C The correlation developed can be used to design an optimal thermal PC system
6.	Mahmoud et al. (2013)	Six PCM Paraffin wax PCM-HS29P – 29°C PCM-HS34P – 34°C PCM-OM37P – 37°C PCM-OM46P – 46°C PCM-HS58P– 58°C Rubitherm RT – 42°C	Six different types of heat sink supplied with power input, $2 \leq Q \leq 6$ W The temperature of heat sinks is lowered with the use of PCM Used for portable electronic components Honeycomb inserts with PCM in heat sink mesh give better results
7.	El Qarnia et al. (2013)	Three PMHS n-eicosane (T_m – 36°C)	2D numerical study supplied with power input per unit length 60 W/m Melting of PCM is due to heat dissipation from the heat sources and the bottom heat source has the lowest temperature

(Continued)

TABLE 2.3 *(Continued)*   Summary of Literature Pertinent to the Phase Change Materials

S. No.	Author and Year	Mode of Heat Transfer and Heat Sources	Key Points
8.	Baby and Balaji (2013)	n-eicosane (T_m – 36.5°C) Paraffin wax (T_m – 53°C – 57°C)	The experimental study was carried out for different pin-fin heat sinks with power input, $5 \leq Q \leq 8$ W ANN + GA combination for the optimal heat sink, volume fraction, and thermal conductivity enhancer (TCE)
9.	Rathod and Banerjee (2014)	Paraffin wax (T_m – 58°C – 60°C)	Experimental investigation on Shell and tube heat exchanger for different parameters 0.0167 kg/s $\leq$ m $\leq$ 0.0833 kg/s 75°C $\leq$ Ti $\leq$ 85°C The melting of PCM at the top end is quick than at its lower end
10.	Ashraf et al. (2017) Ali et al. (2018)	Different geometries of the heat sink Paraffin wax (Tm – 56°C – 58°C) RT-54, RT-44, RT-35HC, SP-31, and n-eicosane	An experimental study to investigate cooling of the heat sink using PCM on various parameters Power input: 4–8 W The highest enhancement ratio was obtained by SP-31 Paraffin wax is suitable for the highest power input of 8 W
11.	Arshad et al. (2018)	Round pin-fin heat sink Paraffin wax (Tm – 56°C – 58°C)	Experiment study to investigate the effect of pin-fin diameter. The following parameter was varied: 2 mm $\leq$ d $\leq$ 5 mm, 1.6 kW/m² $\leq$ q $\leq$ 3.2 kW/m² $0 \leq$ volume of PCM ≤ 1 Heat sink with a 3-mm diameter shows the maximum enhancement ratio
12.	Rehman and Ali (2018)	Heat sink with paraffin wax and copper foam	Experiment investigation on the effect of copper foam porosity on the heat sink temperature 0.95 $\leq$ porosity $\leq$ 0.97, 15 $\leq$ pore density $\leq$ 35 $8 \leq Q \leq 24$ W Better performance is obtained for porosity 0.95 and lower power input
13.	Loganathan and Mani (2018)	Ten PCM Eight criteria were considered	Fuzzy logic modelling The optimum PCM obtained is RT-80 (Tm – 80°C)

(Continued)

TABLE 2.3 *(Continued)*   Summary of Literature Pertinent to the Phase Change Materials

S. No.	Author and Year	Mode of Heat Transfer and Heat Sources	Key Points
14.	Usman et al. (2018)	Heat sink with paraffin wax (T_m – 56°C – 58°C), RT-44 (Tm – 44°C) RT-35HC (Tm – 35°C)	An experimental study to determine the better pin-fin arrangement Staggered and inline arrangements of pin-fin Triangular pin-fin heat sinks Heat supplied $5 \leq Q \leq 8$ W RT-44 showed a maximum enhancement ratio as compared to other PCMs
15.	Zhou et al. (2001)	Three PMHS n-octadecane (Tm – 28°C)	An experimental study was carried out Subcooling and Stefan number play a significant role in the melting of PCM
16.	Zeng et al. (2010)	1-Tetradecanol Silver nanowire	Experiment study to enhance the thermal conductivity of composite PCM
17.	Cui et al. (2011)	Carbon nanofiber, carbon nanotube filled with PCM soy wax and paraffin wax	Experiment study to enhance the thermal conductivity of composite PCM The carbon nanofiber with soy wax was found to be a more effective composite PCM as compared to paraffin wax
18.	Yavari et al. (2011)	Octadecanol and nanostructured graphene	Experiment study to enhance the thermal conductivity of composite PCM The percentage content of the nanostructured PCM was varied
19.	Rabie et al. (2019)	RT35HC (Tm – 41–44)	Experiment and numerical study for different inclination – $45° \leq 0° \leq 45$, $0 \leq$ height ratio $\leq 60\%$ of the heat sink. Heat flux supplied was 3,000 W/m² and the maximum cooling performance was observed at 45°
20.	Khademi et al. (2019)	Oleic acid (Tm – 13°C)	Numerical simulation was carried out with oleic acid and water was used in the enclosure The mixture of PCM with water melts faster and stores maximum energy in extremely cold conditions

(Continued)

TABLE 2.3 *(Continued)* Summary of Literature Pertinent to the Phase Change Materials

S. No.	Author and Year	Mode of Heat Transfer and Heat Sources	Key Points
21.	Gasia et al. (2019)	n-octadecanc (Tm – 27.7°C)	An experimental study was carried out with the use of metal foam and PCM which increased the thermal energy storage by 14% for constant heat transfer fluid temperature and flow rate
22.	Hamza and Mustapha (2019)	Single PMHS n-eicosane with copper oxide particle	2D numerical study was performed and the melting of the PCM was enhanced with the concentration of the nanoparticles
23.	Xie et al. (2019)	Single PMHS PCM-based heat sink Paraffin wax (Tm – 34°C – 36°C)	Numerical study supplied with a heat flux of 50,000 W/m². The topology-optimized tree heat sink showed better thermal performance Heat sink mounted on the heat sources is more significant than the inverted heat sink
24.	Kalbasi et al. (2019)	PCM heat sink RT-27 (Tm – 30°C)	Numerical study to determine the optimum number of fins, height ratio, and volume content of PCM $10 \leq$ height range ≤ 30 mm, $0.2 \leq$ fin thickness ≤ 0.5 mm Heat flux 5,000 and 10,000 W/m² The optimum number of fins was 12, thickness 2.236 mm for 500 W/m²
25.	Bondareva et al. (2019)	Single PMHS n-octadecane Al_2O_3 nanoparticle	Numerical study of PCM-based heat on the angle of orientation and the concentration of the nanoparticles The maximum concentration of the nanoparticles enhances the cooling performance of PCM-based heat sink
26.	Zarma et al. (2019)	Pure salt hydrate (Tm – 29.8°C) nanoparticles (Al2O3, CuO, and SiO_2)	Numerical study The maximum concentration of Al_2O_3 nanoparticles enhances the cooling performance of PCM and the temperature of the solar cells is reduced

(Continued)

TABLE 2.3 *(Continued)*   Summary of Literature Pertinent to the Phase Change Materials

S. No.	Author and Year	Mode of Heat Transfer and Heat Sources	Key Points
27.	Gharbi et al. (2018)	A PCM-based copper heat sink, single PMHS, PCM27 (Tm – 27°C)	Numerical study on the copper-based PCM heat sink Longer fins are more effective than shorter fins (4,000 W/m²)
28.	Zalba et al. (2003)	Review on PCMs	Thermophysical properties, heat transfer characteristics, and analytical and numerical modelling methods were briefed
29.	Mohamed et al. (2009)	Review of PCM for solar energy system	They recommended that PCM with a melting temperature of 50–60°C is appropriate for a solar energy system, which can store the energy during charging and can be utilized while discharging
30.	Hasan et al. (2016)	Salt hydrate (Tm – 32°C) Paraffin wax (Tm – 43°C) Milk fat (Tm – 40°C)	Conducted experiments and numerical investigations for different heat inputs 4 W ≤ Q ≤ 10 W
31.	Saha et al. (2006); Saha and Dutta (2010)	n-eicosane Paraffin wax PCM-based heat sinks	Conducted experiments and numerical investigations for different heat inputs The number of fins of the heat sink was varied The volumetric percentage of TCE was varied and 8% volumetric percentage was found to be better as compared to others
32.	Salunkhe and Shembekar (2012)	Thermal performance of the system	Microencapsulation of the PCM, Characteristics of the PCM was briefed Challenges in implementing microencapsulation PCM were reported
33.	Elmozughi et al. (2014)	NaNO₃-stainless steel encapsulated phase change material (EPCM) capsules	Effects of the thermal expansion and volume expansion on energy storage
34.	Ling et al. (2014)	Battery thermal management system Expanded graphite (EG) paraffin 44°C	Different power input was supplied in a range of 5 W ≤ Q ≤ 15 W Variation of density to study the effect on temperature and latent heat of PCM

(Continued)

TABLE 2.3 *(Continued)*   Summary of Literature Pertinent to the Phase Change Materials

S. No.	Author and Year	Mode of Heat Transfer and Heat Sources	Key Points
35.	Janarthanan and Sagadevan (2015)	Review of PCM for different applications	Reported the use of PCM for different domains and the characteristics of PCM
36.	Kahwaji et al. (2018)	Three different paraffin waxes and three different alkanes	For commercial paraffin waxes $150 \leq$ Latent heat ≤ 200 J/g and greater than 200 J/g for pure alkanes
37.	Hasan and Tbena (2018b)	n-octadecane-27.7 RT44–41 Micro pin-fin heat sink	Numerical investigation with a different PCM, Unfinned heat sink with n-octadecane showed the maximum temperature drop Circular pin-fin PCM heat sink gave a better performance as compared to other geometries
38.	Hasan and Tbena (2018a)	Paraffin wax 56, n-eicosane 36.5, p116–50 and RT41–37.5	Numerically investigated Combination of different PCM placed in micro-channels The lower melting temperature of PCM reduced temperature considerably
39.	Karami and Kamkari (2019)	One-fin and three-fin enclosure lauric acid – 43.5–48	Transient numerical simulation Different inclination angles varied from 0° to 180° Melting time reduces with the inclination angle Minimum for vertical as compared to other orientation > 90°
40.	Song et al. (2019)	Paraffin (41–44) Cold plate	106 cylindrical batteries Cold plate for liquid cooling PCM helps in reducing the surface temperature of the batteries
41.	Khurade et al. (2021a, 2021b)	Paraffin Wax Smart Phone cooling	Used protrude chips Different thermal conductivity substrate boards
42.	Ekbote (2020), Talele et al. (2021)	Review of cooling techniques	Conducted detailed review of cooling techniques. Performed numerical investigation on PCM based mini-channels

2.5 SUMMARY OF THE LITERATURE SURVEY

1. Cooling of the identical discrete IC chips using air under the different heat transfer modes (natural/forced/mixed) was studied.

2. The major focus was on flush and protruding heat sources with 1D and 2D geometry.

3. The numerical analyses adopted for the cooling of the heat sources were finite element analysis (FEA), computational fluid dynamics (CFD) Fluent, finite volume method (FVM), gambit, flotherm, and MATLAB®, and the optimization techniques used were the design of experiments (DOEs), Taguchi method, sequential quadratic programming, and genetic algorithm.

4. Most of the studies dealt with cooling of the heat sources using the PCM-based enclosure, PCM-based heat sinks, and fluid-based micro and mini-channels.

2.6 SCOPE FOR DEVELOPMENT

1. The cooling of the discrete IC chips under the mixed convection heat transfer mode needs to be focused.

2. The non-uniform spacing of the IC chips can be carried out to study the temperature distribution.

3. The cooling studies can be extended to 3D geometry under steady-state mode.

4. Studies on the PCM-based mini-channels are of prime importance to bring down the temperature.

2.7 DIFFERENT PARAMETERS CONSIDERED FOR THE STUDY

The different parameters identified for the present study are mentioned in Table 2.4.

TABLE 2.4   Parameters Selected for the Present Study

S. No.	Parameters	Various Methods Available	Selected Parameters
1.	Cooling technique	Phase change material (PCM) Air cooling/liquid cooling/ micro-/mini-channels	Air cooling + PCM
2.	Modes of heat transfer	Natural convection Forced convection Mixed convection	All modes of heat transfer
3.	Mounting of chips	Flush mounting Protrude mounting	Protrude mounting
4.	Size of chips	1D chips 2D chips 3D chips	3D chips
5.	Geometry of chips	Symmetric Asymmetric	Asymmetric
6.	Arrangement of chips on board	Horizontal Inclined Vertical	Orientation of substrate board 0°, 30°, 60°, and 90°
7.	Analysis of chips	Constant heat rate (q) Constant heat flux (q_f) Constant heat generation (q_v)	Constant heat generation (q_v)
8.	Experimental investigation		
9.	Numerical investigation	ANSYS Icepak (Fluent) Gambit Ansys (FEA) COMSOL Flotherm	ANSYS Fluent
10.	Optimization technique	DOE ANN, GA Fuzzy logic Taguchi method	ANN + GA

CHAPTER 3

Experimental Facility

3.1 INTRODUCTION

This chapter elucidates the details about the experimental facility used for the present study along with the methodology to carry out the experiments under the different heat transfer modes using both air and phase change material (PCM). The uncertainties associated with the measurement of various parameters of the experiment are also explained. To begin with, the chapter highlights the design and selection of the integrated circuit (IC) chips and switch-mode power supply (SMPS) board used for the present analysis. For a better understanding, the chapter is classified into the following sections:

- Selection of the IC chips and SMPS board

- Design of the IC chips and SMPS board

- Experimental facility

- Instrumentation

- Experimental methodology

- Uncertainty analysis

3.2 SELECTION OF THE INTEGRATED CIRCUIT CHIPS AND THE SWITCH-MODE POWER SUPPLY BOARD

The selection of the Switch mode power supply (SMPS) board on which the IC chips are mounted is extremely important for the electronic cooling application. The SMPS board converts the main power using the switching

DOI: 10.1201/9781003188506-3

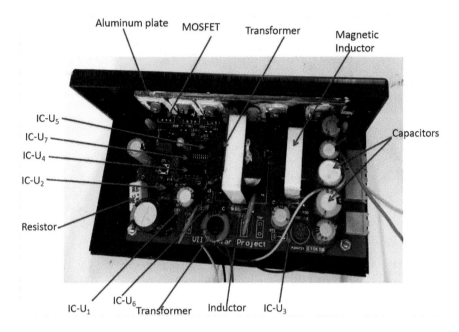

FIGURE 3.1   Different components placed on the SMPS board.

devices and supplies the required voltage or current for the effective working of the IC chips and other components such as an inductor, capacitor, metal oxide semiconductor field effect transistor (MOSFET), and transformer. Figure 3.1 shows the different components mounted on the SMPS board (Mathew and Hotta (2018)).

The SMPS board selected for the present study is designed and developed for the central processing unit (CPU) application powered by solar energy. Figure 3.2 shows the electronic computer-aided design (ECAD) layout of the SMPS board. The ECAD layout is then converted to the mechanical CAD model with the actual size of the components using the ANSYS Icepak (version: R-15), as shown in Figure 3.3. The power rating of each component mounted on the SMPS board is then studied and the seven asymmetric rectangular IC chips (which are the highlights of the present study) are identified. The dimensions of all the seven IC chips (designated as U) and the SMPS board (referred to as substrate board) are given in Table 3.1. The power rating and size of the selected IC chips are fetched from the datasheet of each IC chip, respectively ((Texas Instruments, 2002; 2014) and Vishay (2002)). The selected IC chips are designated as U_1–U_7, as shown in Figure 3.4.

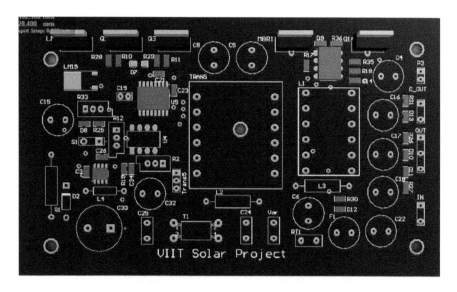

FIGURE 3.2   Electrical CAD design of the SMPS board.

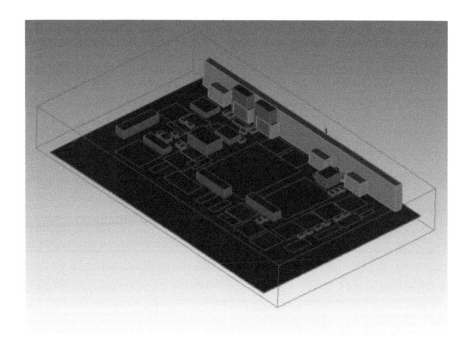

FIGURE 3.3   Mechanical CAD layout of the SMPS board.

TABLE 3.1   Specifications of Different Components Mounted on the SMPS Board

Components	Dimensions ($l_c \times w_c \times t_c$), cm	Heat flux (q), W/cm^2
U_1	$1.016 \times 0.76 \times 0.265$	2.518
U_2	$0.5 \times 0.4 \times 0.175$	5.46
U_3	$1.016 \times 0.762 \times 0.355$	0.058
U_4	$1.016 \times 0.762 \times 0.355$	0.058
U_5	$1.05 \times 0.76 \times 0.265$	2.005
U_6	$0.5 \times 0.4 \times 0.175$	4.15
U_7	$1.965 \times 0.65 \times 0.343$	0.082
Substrate board	$15.24 \times 10.16 \times 0.164$	NA
Enclosure	$30 \times 15 \times 10$	NA

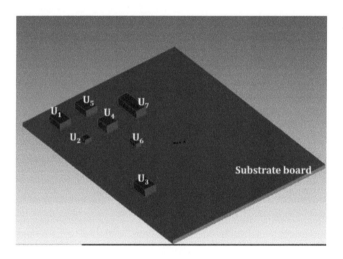

FIGURE 3.4   Final layout of SMPS board with IC chips.

3.3 DESIGN OF THE INTEGRATED CIRCUIT CHIP AND THE SWITCH-MODE POWER SUPPLY BOARD

3.3.1 Design of Integrated Circuit Chips

The aluminium blocks are used to mimic the actual IC chips and are called heat sources. The power input (heat input) to the IC chips is facilitated using a heater with 80/20 coil-type Nichrome wire (80% nickel and 20% chromium). The heater wires (covered with Teflon tape for insulation) are placed inside the groove made on the IC chip's bottom face, as shown in Figure 3.5. The edge thickness for all the sides at the bottom face of each IC chip is maintained uniform as 1 mm. The slots are provided at the bottom

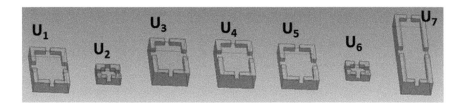

FIGURE 3.5   Cavity on the bottom face of each IC chip.

face of each IC chip for placing the thermocouple wires and heater wires which are further connected to a DC power source. A mixture of the thermal paste and Araldite is used to keep the thermocouple wires and heater wires intact inside the IC chip cavity. The layout of a particular IC chip U_7 is shown in Figure 3.6. The final layout of the IC chip with the thermocouple and the heater wires is shown in Figure 3.7.

The K-type thermocouples are employed to measure the temperature of the IC chips. A rectangular slot is provided on the edges of all the IC chips to place the beaded thermocouples. Two more slots are made on the IC chips to place the heater wires which are further connected to a DC power source. The ice point and boiling point of water are the standards used to calibrate each thermocouple using a standard mercury thermometer. The

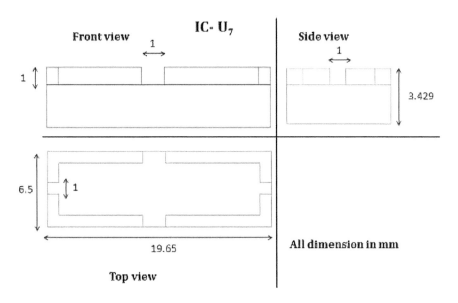

FIGURE 3.6   Design of the IC chip U_7.

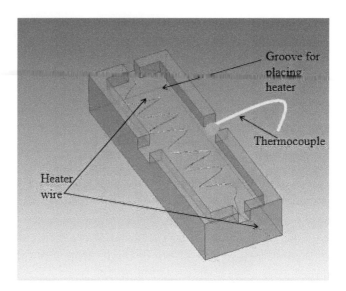

FIGURE 3.7   Final layout of the IC chip with thermocouple and heater wires.

temperature measurement error is found to be ± 0.2°C. The other end of the thermocouples is connected to the temperature logger (Make: UniPro Log Plus). One thermocouple is placed at the bottom of each IC chip and two are on the substrate board to measure the board temperature (Mathew and Hotta (2021a)).

3.3.2 Design of the Switch-Mode Power Supply (Substrate) Board

The substrate board is divided into seven rows and five columns (7×5) for arranging the seven asymmetric IC chips on it. The dimensions of the substrate board showing the 35 positions (11–75) are shown in Figure 3.8. Out of the seven IC chips, three are dissipating very low power, i.e. U_3 (0.058 W), U_4 (0.058 W), and U_7 (0.082 W) as compared to the other IC chips (as given in Table 3.1), due to which their temperature is found to be very low (close to ambient, 25–27°C); hence, these three IC chips do not take part in thermal interaction with the remaining four high-power IC chips (U_1, U_2, U_5, and U_6). Their positions are fixed on the substrate board at 55 (U_3)-24 (U_4)-23(U_7), respectively. The fixed positions are highlighted with gray in Figure 3.8. These are the original positions of the IC chips (U_3, U_4, and U_7) on the SMPS board as shown in Figure 3.4. Therefore, now only 32 positions are available for arranging the other four IC chips on the board, and they can be arranged in 863040 possible ways ($^{32}P_4$). The IC chips U_2, U_4, and U_5

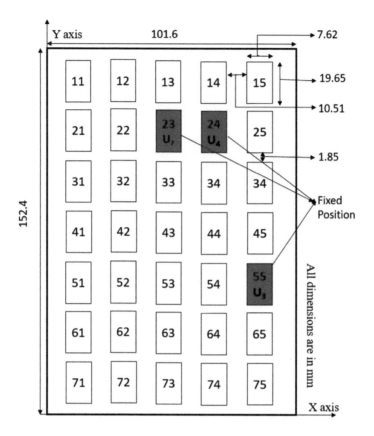

FIGURE 3.8 Dimensions of the substrate board, showing the 35 positions for arranging the seven IC chips.

are at the top surface of the SMPS board, while the IC chips U_1, U_3, U_6, and U_7 are at the bottom surface of the board, as shown in Figure 3.1.

3.3.2.1 Substrate Board Design to Carry Out the Laminar Forced Convection Experiments

The substrate pieces hold the IC chips (there are 35 substrate pieces from 35 slots of the board) along with the assembly of the thermocouple and heater wires, as shown in Figure 3.7. Therefore, the seven IC chips can be accommodated in the seven positions on the substrate board and the left-out substrate pieces are used to fill the remaining 28 positions, as shown in Figure 3.9. The final substrate board on which the IC chips are mounted is fitted at the centre of the test section of the experimental facility (details

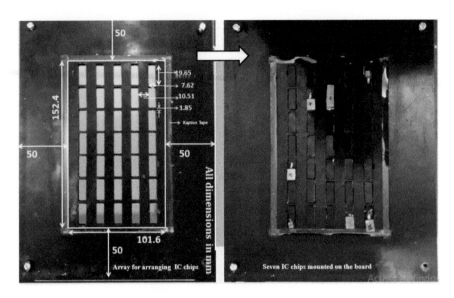

FIGURE 3.9   Final layout of the substrate board.

are explained under Section 3.7), as shown in Figure 3.10. The circular shaft of the frame is connected to the bevel gear arrangement which has the protector and pointer at the top and a lever at the bottom for facilitating the rotation. The substrate board is rotated at different angles of 0°, 30°, 60°, and 90° with the help of the lever, as shown in Figure 3.11.

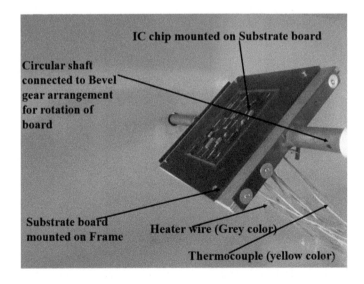

FIGURE 3.10   Schematic arrangement of the substrate board rotation.

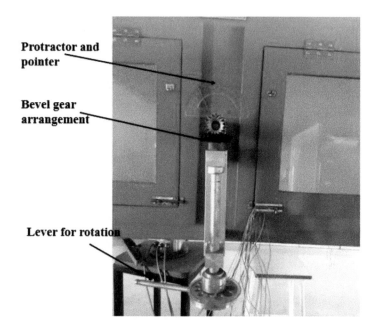

Protractor and pointer

Bevel gear arrangement

Lever for rotation

FIGURE 3.11   Bevel gear arrangement for substrate board rotation.

3.3.2.2 Substrate Board Design to Carry Out the Experiments Using the Phase Change Material-Filled Mini-Channels

Aluminium mini-channels are manufactured as per the dimensions shown in Figure 3.12. These channels are designed and placed on the substrate board adjacent to the IC chips. The channels are filled with the PCMs to enhance the working cycle of the IC chips. The commercially available Paraffin wax is selected as the PCM for the current study whose melting point temperature ranges from 48.28–54.21°C. The PCM is available in solid form at room temperature and is then crushed into powder (as shown in Figure 3.13) to be filled inside the mini-channels. Two thermocouples are inserted at the edge of each channel to measure the channel temperatures and two are immersed inside the PCM to measure their temperature. The interface between the surface of the IC chips and the channels is covered with a thermal paste to reduce the air resistance and to provide maximum contact surface between the IC chips and the mini-channels. The final assembly of the substrate board with the PCM-filled mini-channels along with the thermocouple wires is shown in Figure 3.14.

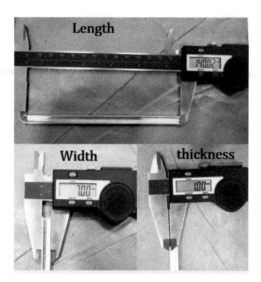

FIGURE 3.12   Dimensions of the mini-channels used for the present study.

The PCM melting point temperature is determined by placing a thermocouple inside it and also by the characterization technique using the differential scanning calorimeter (DSC) (National Chemical Laboratory, Pune). Figure 3.15 shows the characterization curve of the Paraffin wax. Table 3.2 gives its thermo-physical properties both in solid and liquid forms (*Fuels and Lubricants Handbook*, Totten et al. (2003)).

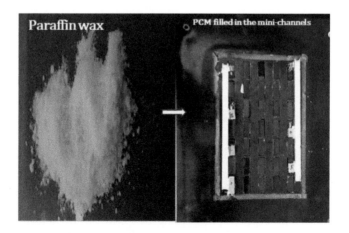

FIGURE 3.13   Mini-channels filled with PCM.

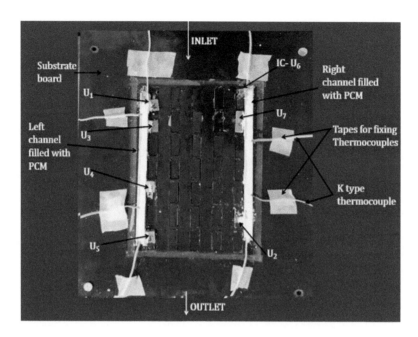

FIGURE 3.14   Final layout of the substrate board with PCM-filled mini-channels.

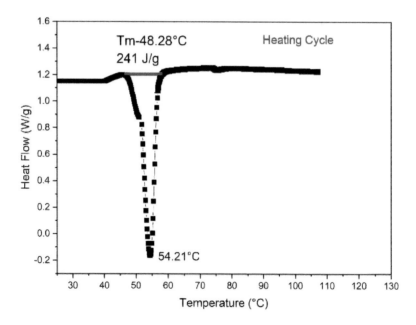

FIGURE 3.15   Characterization curve of paraffin wax using the differential scanning calorimetry (DSC).

TABLE 3.2   Properties of the Paraffin Wax

Sl. No.	Property	Value
1.	Density, kg/m³	818 (s), 760 (l)
2.	Thermal conductivity, W/mK	0.24 (s), 0.22 (l)
3.	Dynamic viscosity, Ns/m²	0.0342 (l)
4.	Specific heat, J/kgK	2,950 (s), 2,510 (l)
5.	Latent heat of fusion, kJ/kg	266 (both s and l)
6.	Melting point, °C	48.28–54.21 (melting range)

3.4 EXPERIMENTAL SETUP AND INSTRUMENTATION

The low-speed horizontal wind tunnel is the main component of the present experimental facility and consists of the effuser (inlet duct), test section, diffuser, axial flow fan, and motor with the speed control unit. The effuser consists of the honeycomb section to guide and straighten the airflow to the test section which helps in reducing the axial and lateral turbulence stresses. The mesh size and length of the honeycomb section are considered as per the recommended standards. The effuser has a cross section of 1,800 mm × 1,800 mm with a contraction ratio of 9:1 connected to the test section of cross section 600 mm × 600 mm. The other end of the test section is connected to a diffuser with flanges which enlarges to a diameter of 1,200 mm. A fan is the independent unit of the wind tunnel and is housed at the circular edge of the diffuser. The fan speed is controlled using the thermistor connected to the three-phase motor with a 440 AC supply. All the safety precautions for excessive electrical loading are also provided for the wind tunnel. The wind tunnel is made of wood to maintain the adiabatic condition all around it. The photograph view of the wind tunnel is shown in Figure 3.16. The test section has a cross section of 60 cm × 60 cm with a length of 2 m. A provision is made in the test section for mounting the substrate board at the centre, and the board is then rotated at various angles using the bevel gear arrangement, as shown in Figure 3.11. The substrate board (Bakelite) is mounted on the frame which is located at the test section centre and rotated for different angles 0°, 30°, 60°, and 90°, respectively, as shown in Figure 3.17. The airflow to the test section is induced by a fan located at the diffuser and is rotated at different revolutions per minute (RPMs) to vary the air velocity.

3.4.1 Instruments Used for the Experimental Analysis

The different instruments used for the analysis of the experiments are the DC power source, temperature data logger, hot wire anemometer, and digital multimeter. These are explained in the subsequent sections.

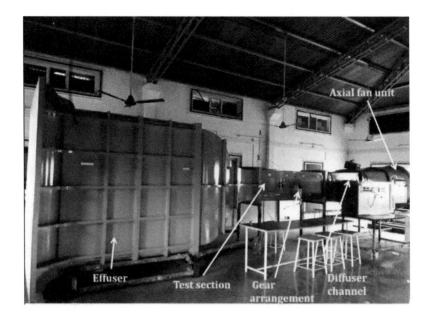

FIGURE 3.16   Pictorial view of the wind tunnel.

3.4.1.1 Direct Current Power Source

The heat input to the seven IC chips is independently controlled using the three dual DC power sources as shown in Figure 3.18. The DC power source (Make: MULTISCOPE) is a dual output with voltage and current ranges of 0–30 V and 0–2 A, respectively, and is designed to meet industrial and academic standards. It is provided with a trip function facility

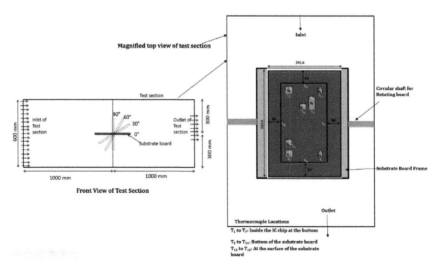

FIGURE 3.17   2D view of the test section showing the substrate board rotation.

FIGURE 3.18   Pictorial view of the DC power source.

to avoid any damage to the electronic circuit during the testing and the power supply. The technical details of the DC power source are provided in Table 3.3. The heat input to the five IC chips (U_1, U_2, U_5, U_6, and U_7) is supplied independently from the DC power source. However, for the other two IC chips (U_3 and U_4), the heat is supplied by connecting the power sources in parallel, as both these chips dissipate the same power, as given in Table 3.1.

3.4.1.2 Hot Wire Anemometer

The air velocity in the present study is measured using a hot-wire anemometer, as shown in Figure 3.19. It consists of a telescopic probe (Make: Lutron AM-4204) of diameter 12 mm and its length can be extended up to 940 mm to measure the velocity at any location inside the test section. The telescopic probe can measure the velocity in the range of 0.2–20 m/s and operates in the temperature range of 0–50°C. The sampling time of the anemometer is approximately 0.8 s with the accuracy of the velocity measurement as ±1% of full-scale reading.

TABLE 3.3   Specification of the DC Power Source

Parameter	Range
Voltage output	0–30 V, DC
Current output	0–2 A
Voltage resolution	10 mV
Current resolution	1 mA
Supply voltage	230 V, 50 HZ

FIGURE 3.19   Pictorial view of the hotwire anemometer.

3.4.1.3 Temperature Data-Logger

The temperatures of the IC chips are recorded using the data acquisition system (DAQ) (Make: UniPro Log Plus) using the K-type thermocouples, as shown in Figure 3.20. It has a total of 16 channels in one module which is connected to the computer to record and store the temperature data. The DAQ can measure voltage, current, resistance, and period apart from the temperature. Each channel has four ports, which are used for the temperature measurement through the resistance temperature detector (RTD) and thermocouples, also for the voltage and current measurement. The thermocouples have a fixed resolution of 0.1°C, a scanning rate of 500 ms per channel, and an accuracy of ±0.25%. One edge of the thermocouple is

FIGURE 3.20   Pictorial view of the data acquisition system.

FIGURE 3.21   Pictorial view of the multimeter.

connected to the IC chips and another end is connected to the DAQ. The DAQ is connected to the computer for storing the data through the RS485 serial communication port (SCP).

3.4.1.4 Digital Multimeter

A digital multimeter (Make: HTC DM97), as shown in Figure 3.21, is used to verify the readings obtained with the DC power source. The multimeter is connected in series and parallel with the heater for the current and voltage measurement, respectively. It has a voltage and current resolution of 0.1 mV and 0.1 μA, respectively.

3.4.1.5 Kapton Tape

The Kapton tape, as shown in Figure 3.14, is used to cover the slots on the periphery of the original substrate board and to prevent the air from entering into the slots. This removes eddies formation to enhance the thermal performance of the IC chips.

3.5 EXPERIMENTAL METHODOLOGY

The methodology for the experimental analysis on seven IC chips positioned at different places on a substrate board and supplied with non-uniform heat fluxes is detailed further. The experiments are carried out in two phases; phase I deals with the laminar forced convection steady-state

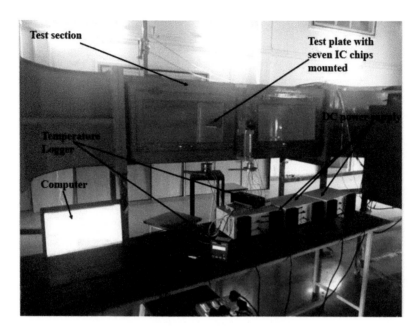

FIGURE 3.22 Experimental facility used for the present study.

heat transfer analysis from the IC chips for different substrate board orientations (explained under Section 3.5.1), and phase II deals with transient analysis on phase change material-filled mini-channels under the natural convection heat transfer mode (explained under Section 3.5.2). The experimental test rig is shown in Figure 3.22.

3.5.1 Procedure for Conducting Laminar Forced Convection Steady-State Experiments

The procedure for conducting the laminar forced convection steady-state experiments is explained as follows:

1. Initially, the test section (substrate board with IC chips) is supplied with uniform air velocity (through the axial flow unit of the wind tunnel) to maintain the constant domain temperature, i.e. very close to the ambient (25–30°C).

2. The required heat input to the IC chips is set by switching ON the DC power source and by adjusting the required voltage and current.

3. The scanning of the temperature logger is initiated through the SCP and then the temperature of the IC chips from each thermocouple

is stored in the computer after reaching the steady-state condition (there should not have been a temperature variation of more than ±0.1°C). The voltage and current from the DC power source are also recorded.

4. The temperature excess (temperature difference between the IC chip and the ambient) for each IC chip is then calculated using Equation 3.1.

$$T_{excess} = T_{hs} - T_\infty \tag{3.1}$$

5. The conduction heat loss from the IC chips to the substrate board, insulation heat loss, and the radiation heat loss from the IC chips to the ambient are calculated using Equations 3.3–3.5, respectively, given under Section 3.6.1.

6. The contribution of convection (leading to the cooling of IC chips) is then evaluated using Equations 3.6 and 3.7, respectively given under Section 3.6.1. A comparison is then made for all the seven IC chips.

7. The non-dimensional numbers affecting the laminar forced convection heat transfer for the IC chips are then evaluated using Equations 3.8 and 3.9, respectively, given under Section 3.6.1.

8. Steps 1–7 are repeated for different substrate board orientations (0° (horizontal), 30°, 60°, and 90° (vertical)), for different air velocities (4.5 m/s and 8 m/s), and for different input heat fluxes to the IC chips as mentioned in Table 3.4.

TABLE 3.4   Specifications of the IC Chips with Their Input Heat Fluxes under Laminar Forced Convection

Dimensions ($l_c \times w_c \times t_c$), cm	Dimensions ($l_c \times w_c \times t_c$), cm	Heat flux (q), W/cm²
U_1	$1.045 \times 0.76 \times 0.265$	2.518
U_2	$0.5 \times 0.4 \times 0.175$	5.46
U_3	$1.016 \times 0.762 \times 0.355$	1.9375
U_4	$1.016 \times 0.762 \times 0.355$	1.9375
U_5	$1.05 \times 0.76 \times 0.265$	2.005
U_6	$0.5 \times 0.4 \times 0.175$	4.15
U_7	$1.965 \times 0.65 \times 0.343$	1.1743
Test section	$200 \times 60 \times 60$	NA
Original substrate board	$15.24 \times 10.16 \times 0.6$	NA
Extended substrate board	$25.24 \times 20.16 \times 0.6$	NA

3.5.2 Procedure for Conducting Transient Experiments on the Phase Change Material-Filled Mini-Channels under the Natural Convection

The procedure for conducting the transient experiments on PCM-filled mini-channels under the natural convection heat transfer mode is explained next.

1. The final assembly of the PCM-filled mini-channels (as shown in Figure 3.14) is kept inside the test section with a horizontal substrate board. All the experiments for the PCM-filled mini-channels are conducted only for the horizontal board.

2. The IC chips are supplied with four different cases of volumetric heat generation (leading to non-uniform heat fluxes) as mentioned in Table 3.5 (Mathew and Hotta (2021b)).

3. Transient experiments are then conducted using the mini-channels with and without the PCM.

4. The experiments are conducted for two set point parameters (SPPs); the first parameter is the time taken by the IC chips to reach the set point temperature (SPT) less than 100°C without the PCM-based mini-channels (WPMC), and the second parameter is the time taken by the PCM to reach the 90% melting (MPMC) so that the liquid PCM does not flow outside the channel, as the channels are open at both the ends.

5. The IC chips are placed beside the PCM-filled mini-channels and it is observed that the left PCM-based mini-channel (LPMC) reaches 90% melting faster. Therefore, the power supply to the IC chips U_1, U_3, U_4, and U_5 are cut off, and the power supply to the remaining IC

TABLE 3.5   Variable Heat Inputs to the IC Chips for the PCM Filled Mini-Channel Cases

IC chips	Case 1 (1×10^7, W/m³)	Case 2 (0.8×10^7, W/m³)	Case 3 (0.6×10^7, W/m³)	Case 4 (0.4×10^7, W/m³)
U_1	2.1	1.68	1.26	0.84
U_2	0.35	0.28	0.21	0.14
U_3	2.75	2.2	1.65	1.1
U_4	2.75	2.2	1.65	1.1
U_5	2.11	1.69	1.27	0.85
U_6	0.35	0.28	0.21	0.14
U_7	4.38	3.5	2.63	1.75

chips (U_2, U_6, and U_7) is kept ON till the PCM in the right PCM-based mini-channel (RPMC) reaches 90% melting.

6. The temperatures of the IC chips, mini-channels, and PCM from the temperature logger are then recorded on the computer.

7. The conduction heat loss from the IC chips to the substrate board, IC chips to the mini-channel, insulation heat loss from the IC chips, and the radiation heat loss from the IC chips to the ambient are calculated using Equations 3.11–3.14, respectively, given under Section 3.6.2.

8. The contribution of convection (leading to the cooling of IC chips) is then evaluated using Equations 3.15 and 3.16, respectively, given under Section 3.6.2. A comparison is also made for all seven IC chips.

9. The non-dimensional numbers affecting the natural convection heat transfer for the IC chips are then evaluated using Equations 3.17 and 3.18, respectively, given under Section 3.6.2.

10. The steps 1–8 are then repeated for the different mini-channel cases.

3.6 EXPERIMENTAL CALCULATIONS

The following assumptions are considered for performing the calculations for the present experimental analysis:

1. The entire unit of the wind tunnel is made of wood and is considered to be adiabatic.

2. As the substrate board is longer (200 mm) than its thickness (6 mm), hence, the conduction losses across its depth are significant and the losses along its length are neglected.

3. The IC chips (aluminium) are assumed to be isothermal, as the chip sizes are very small and, hence, the temperature difference at various chip locations is negligible.

4. Uniform melting of the PCM inside the mini-channels is considered neglecting the volume variation due to the PCM melting under the natural convection.

5. The transient experiments on the PCM-based mini-channels under the natural convection are conducted by keeping the test plate at the centre of the wind tunnel test section. As for the test section and the IC chips,

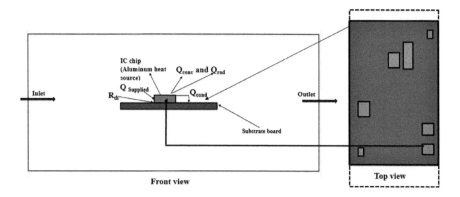

FIGURE 3.23   Energy balance of an IC chip considered for the present analysis.

heights are 300 mm and 2 mm, respectively; hence, this does not affect the formation of the thermal boundary layer on the IC chips.

3.6.1 Experimental Calculations under Laminar-Forced Convection Heat Transfer Mode

Figure 3.23 shows the energy balance of a particular IC chip considered for the present analysis. The energy balance equations considered for the experimental analysis under the laminar forced convection are given in Equations 3.2–3.9.

$$Q_{supp} = V_s I \tag{3.2}$$

$$Q_{cond} = k_{sub} A_{sub} (T_{IC} - T_{sub}) / t_{sub} \tag{3.3}$$

$$Q_{insu} = k_{insu} A_{insu} (T_{IC} - T_{insu}) / t_{insu} \tag{3.4}$$

$$Q_{rad} = F \varepsilon \sigma A (T_{IC}^4 - T_{\infty}^4) \tag{3.5}$$

$$Q_{conv} = Q_{supp} - (Q_{cond} + Q_{insu} + Q_{rad}) \tag{3.6}$$

$$h_{conv} = Q_{conv} / A_{conv} (T_{IC} - T_{\infty}) \tag{3.7}$$

$$Nu_{conv} = h_{conv} L_c / k_f \tag{3.8}$$

$$Re = V L_c / \nu \tag{3.9}$$

Equation 3.2 is the heat input to the IC chips. Equations 3.3 and 3.4 are the conduction and insulation heat losses from the IC chips to the substrate board and insulation, respectively. Here "A_{sub}" and "A_{insu}" are the surface areas of the IC chips and the insulation which are in contact with the substrate, respectively. Equation 3.5 is the radiation heat loss from the IC chips to the ambient, where "A" is the IC chip's surface area. Equation 3.6 is the energy balance of the system and Equation 3.7 is the convection contribution, where "A_{conv}" is the area of all the IC chip faces that are exposed to ambient.

3.6.2 Experimental Calculations for the Phase Change Material-Filled Mini-Channels under the Natural Convection Heat Transfer Mode

The energy balance equations considered for the experimental analysis for the PCM-based mini-channels under the natural convection are given in Equations 3.10–3.18. Equation 3.10 gives the heat input to the IC chips. Equations 3.11 and 3.12 are the conduction heat losses from the IC chips to the substrate board and mini-channel, respectively. Here "A_{sub}" and "$A_{channel}$" are the surface area of the IC chips which are in contact with the substrate board and mini-channel, respectively. Equation 3.13 gives the insulation heat loss from the IC chips to the cork insulation, where "A_{insu}" is the area of the cork insulation in contact with the substrate board. Equation 3.14 gives the radiation heat loss from the IC chips to the ambient, where "A" is the surface area of the IC chips. Equation 3.15 is the energy balance of the system and Equation 3.16 gives the contribution of convection (leading to a cooling of the IC chips), where "A_{conv}" is the area of all the four faces of the IC chips that are exposed to the ambient.

3.7 ERROR ANALYSIS

The uncertainty analysis for both the primary and derived quantities involved in the experiment is carried out. The uncertainties involved in the measurement of the primary quantities are obtained by the calibration of the instrument with which the quantity is measured with a standard instrument, which is found to be accurate. For example, the voltage and current readings of the DC power source are calibrated using a digital multimeter and the thermocouples are calibrated using a mercury thermometer. The uncertainties of derived quantities are calculated based on the uncertainties of the primary quantities and using the uncertainty formula given in Equation 3.19 (Venkateshan, 2008). The uncertainty values

TABLE 3.6   Uncertainty Involved in the Physical Quantities

Sl. No.	Measured quantity	Uncertainty in measurement	Unit
1.	Current (measured)	±0.002 (full scale)	A
2.	Temperature (measured)	±0.2 (full scale)	°C
3.	Voltage (measured)	±0.05 (full scale)	V
4.	Power input (derived)	±0.0583 (full scale)	W
5.	Heat transfer coefficient (derived)	±0.00153	W/m²K

of the primary and derived quantities involved in the experiment are given in Table 3.6. The uncertainty calculation for the power input is given in Appendix D.

Here, m and σ are the primary (measured) and derived quantities, respectively, and Δm and Δσ are the error involved in the primary and derived quantities, respectively.

CHAPTER 4

Hybrid Optimization Strategy for the Arrangement of IC Chips under the Mixed Convection

4.1 INTRODUCTION

Mixed convection is ideally suited for the air cooling of the IC chips under low to medium heat flux ranges. This chapter highlights the numerical analysis on seven asymmetric IC chips oriented at various positions on a horizontal substrate board cooled using air under the mixed convection heat transfer mode, to decide their optimal configuration. For a better understanding, the chapter is classified into the following sections:

- Introduction to the non-dimensional geometric distance parameter (λ)

- Numerical framework

- Temperature variation of the seven IC chips mounted on the switch-mode power supply (SMPS) board

DOI: 10.1201/9781003188506-4

- Optimal arrangement of seven IC chips using the hybrid optimization technique

4.2 NON-DIMENSIONAL GEOMETRIC DISTANCE PARAMETER (λ)

The design of the substrate board is explained under Section 3.3.2. The seven asymmetric rectangular IC chips can be arranged in 863,040 ways (leading to different configurations) on the SMPS board. Each configuration is characterized by a unique non-dimensional geometric distance parameter, λ, and for the present study, λ varies from 0.25103 to 1.87025. A MATLAB® code (the detailed code is given in Appendix B) is used to determine the different configurations of the IC chips along with their corresponding λ values. The λ depends strongly on the positions of the IC chips on the substrate board and their size and is calculated using Equation 4.1.

$$\lambda = \frac{\sum_{i=1}^{7} d_i^2}{l^2 + Y_c^2} \tag{4.1}$$

The subscript "i" is the IC chip number. In Equation 4.1, d_i^2 is $\sum(Xi - Xc)^2 + \sum(Yi - Yc)^2$, where (Xi, Yi) are the centroids of the ith heat source measured from the x- and y-axes of the substrate board, respectively, and (X_c, Y_c) denotes the centroid of the configuration. The centroid of each IC chip and the configuration is shown in Figure 4.1. The l is the total working area dimension of the substrate board measured from the y-axis, and for the present case, it is 101.6 mm.

For example, for the configuration 14-34-55-24-15-25-23 shown in Figure 4.1, the X_i values of the seven IC chips are 68.70, 66.90, 86.84, 68.71, 86.83, 85.03, and 50.02 mm, respectively, and the corresponding Y_i values are 136.075, 90.35, 49.93, 114.43, 136.10, 111.85, and 119.75 mm, respectively. So, the X_c, Y_c, l^2, and d d_i^2 values are calculated as 70.27 mm, 111.37 mm, 10,322.56 mm², and 6,702.26 mm², respectively; thus, the λ value for this configuration is found to be 0.29490. Similarly, the λ calculation is carried out for all the 863,040 configurations. For the present numerical study, 47 different configurations are randomly selected from the whole range of λ values in such a way that they lie between the lower extreme (0.25103) and upper extreme (1.87025) λ values. Four different configurations of the IC chips are shown in Figure 4.2.

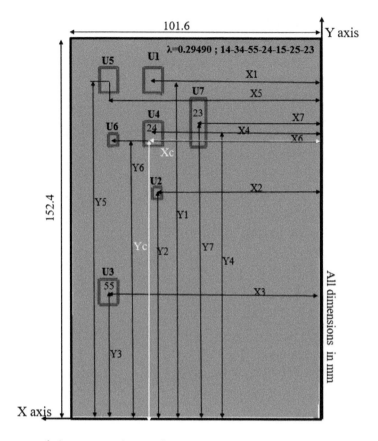

FIGURE 4.1 λ for a particular configuration.

4.3 NUMERICAL FRAMEWORK

The numerical simulations are carried out on the seven asymmetric rectangular IC chips mounted at different positions on an SMPS board and kept inside an enclosure. The simulations are performed using ANSYS Icepak V16.0 under the mixed convection heat transfer mode using air, where the IC chips are supplied with non-uniform heat fluxes. This software is specially designed to be used in the electronic cooling industry. Icepak has the capability of modelling the electronic components and is widely used in the electronic industries for analyzing the heat dissipation rate from the components. The advantage of the Icepak is that all the electronic components are available in its library, i.e. from enclosure to different IC packages like Field Programmable Gate Array (FPGA), Quad Flat No-lead (QFN), fan, and heat exchanger, etc. The other commercial

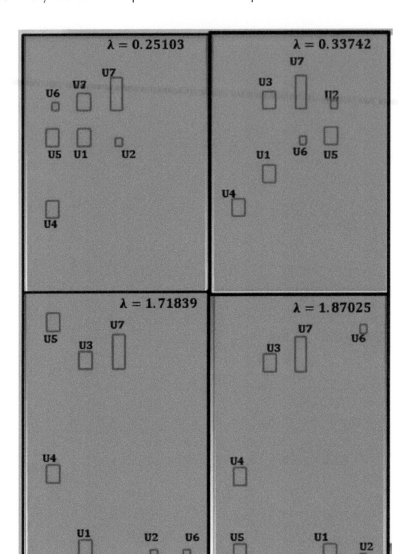

FIGURE 4.2 Four different configurations of λ.

software available for electronic cooling is Mentor graphics – FloTHERM (Siemens), 6 sigma ET – Future, etc. In the present study, the computational model shown in Figure 4.3 consists of an enclosure, substrate board (made of FR_4) with inlet and outlet opening, seven asymmetric IC chips (made of aluminium). The enclosure dimension is taken concerning the actual mounting of the SMPS board.

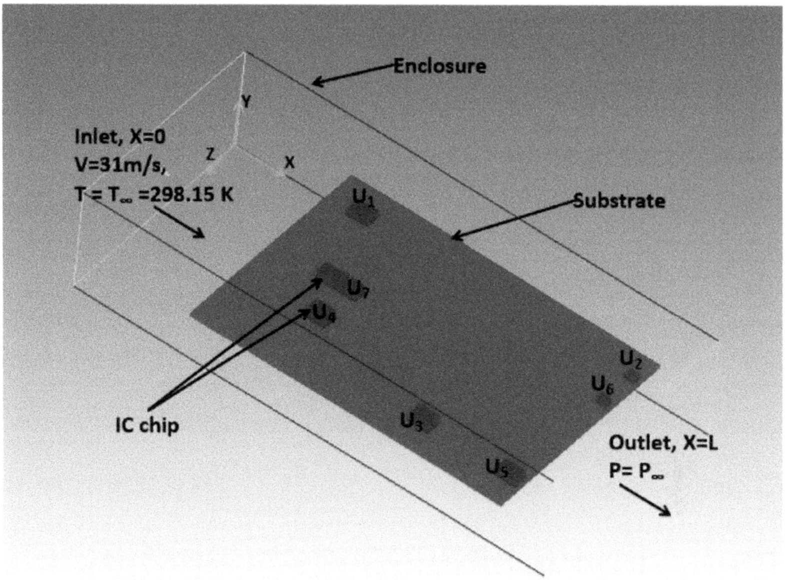

FIGURE 4.3   Computational model used for the present analysis.

The numerical simulations are performed for the 47 configurations (selected out of 863,040 configurations) by supplying the heat flux values as mentioned in Table 3.1. The 47 configurations are selected by covering the lower extreme ($\lambda = 0.25103$) and upper extreme ($\lambda = 1.87025$) values, and the other λ values are spread across equal intervals between the lower and upper extreme ones. The 47 configurations along with their λ values are given in Table 4.1. The low-powered IC chips U_3, U_4, and U_7 having fixed positions on the board (shown in Figure 4.1) are also considered for the simulation.

4.3.1 Governing Equations

ANSYS Icepak uses the Fluent computational fluid dynamics (CFD) solver for the thermal and fluid flow characterization. It uses the fundamental governing equations given in Equations 4.2–4.6. In the present case, 3D steady-state numerical simulations are carried out. The 3D steady-state mass conservation equation (the continuity equation) is expressed as given in Equation 4.2.

The continuity equation for steady-state is expressed as given in Equation 4.3.

$$\frac{\partial u}{\partial x} + \frac{\partial v}{\partial y} + \frac{\partial z}{\partial w} = 0 \tag{4.2}$$

TABLE 4.1   Various IC Chip Configurations Considered for the Numerical Analysis

λ	U_1-U_2-U_3-U_4-U_5-U_6-U_7	λ	U_1-U_2-U_3-U_4-U_5-U_6-U_7
0.25103	34-33-55-24-35-25-23	1.11225	64-45-55-24-61-75-23
0.27076	34-35-55-24-45-25-23	1.12322	35-32-55-24-72-75-23
0.29490	14-34-55-24-15-25-23	1.14604	35-64-55-24-63-71-23
0.33261	13-15-55-24-14-33-23	1.17172	41-62-55-24-14-75-23
0.33742	44-22-55-24-32-34-23	1.20725	75-52-55-24-11-51-23
0.36978	21-32-55-24-22-34-23	1.22138	25-31-55-24-75-61-23
0.39832	42-12-55-24-34-33-23	1.27445	73-34-55-24-11-75-23
0.50345	41-13-55-24-25-14-23	1.30112	72-12-55-24-13-73-23
0.57550	43-44-55-24-52-64-23	1.37202	14-72-55-24-63-74-23
0.59576	31-14-55-24-21-52-23	1.39440	53-71-55-24-15-73-23
0.60495	53-21-55-24-32-15-23	1.45328	75-62-55-24-73-15-23
0.65556	25-22-55-24-12-64-23	1.47204	12-75-55-24-62-72-23
0.67502	54-25-55-24-31-11-23	1.49456	72-41-55-24-71-11-23
0.72940	13-31-55-24-52-51-23	1.50559	65-73-55-24-14-71-23
0.76511	34-32-55-24-71-33-23	1.52132	21-74-55-24-73-75-23
0.79245	34-33-55-24-61-65-23	1.55422	75-72-55-24-14-74-23
0.82344	12-34-55-24-72-44-23	1.59015	15-71-55-24-75-62-23
0.86305	25-41-55-24-34-75-23	1.66312	71-74-55-24-15-75-23
0.88365	51-35-55-24-32-73-23	1.70204	15-73-55-24-75-71-23
0.95786	21-43-55-24-25-71-23	1.71783	74-72-55-24-15-71-23
1.02674	65-42-55-24-71-53-23	1.78906	75-71-55-24-11-72-23
1.03299	71-44-55-24-54-15-23	1.78947	71-75-55-24-73-11-23
1.09816	42-14-55-24-75-62-23	1.86755	74-71-55-24-74-11-23
1.87025	72-71-55-24-75-11-23		

The steady-state momentum equation in an inertial reference frame is given in Equations 4.3–4.5.

$$\left(\frac{\partial u}{\partial t} + u\frac{\partial u}{\partial x} + v\frac{\partial u}{\partial y} + w\frac{\partial u}{\partial z} \right) = -\frac{1}{\rho}\frac{\partial P}{\partial x} + \upsilon\left(\frac{\partial^2 u}{\partial x^2} + \frac{\partial^2 u}{\partial y^2} + \frac{\partial^2 u}{\partial z^2} \right) + \frac{g\beta(T_{IC} - T_\infty)L_c}{u^2} \quad (4.3)$$

$$\left(\frac{\partial v}{\partial t} + u\frac{\partial v}{\partial x} + v\frac{\partial v}{\partial y} + w\frac{\partial w}{\partial z} \right) = -\frac{1}{\rho}\frac{\partial P}{\partial y} + \upsilon\left(\frac{\partial^2 v}{\partial x^2} + \frac{\partial^2 v}{\partial y^2} + \frac{\partial^2 v}{\partial z^2} \right) \quad (4.4)$$

$$\left(\frac{\partial w}{\partial t} + u\frac{\partial w}{\partial x} + v\frac{\partial w}{\partial y} + w\frac{\partial w}{\partial z} \right) = -\frac{1}{\rho}\frac{\partial P}{\partial z} + \upsilon\left(\frac{\partial^2 w}{\partial x^2} + \frac{\partial^2 w}{\partial y^2} + \frac{\partial^2 w}{\partial z^2} \right) \quad (4.5)$$

Equations 4.3–4.5 are derived from Newton's second law of motion and describe the conservation of momentum in the fluid flow. These are Navier-Stokes equations.

The steady-state energy conservation equation is given in Equation 4.6. For IC chips

$$\left(\frac{\partial T}{\partial t} + u\frac{\partial T}{\partial x} + v\frac{\partial T}{\partial y} + w\frac{\partial T}{\partial z} \right) = \frac{k}{\rho C_p}\left(\frac{\partial^2 T}{\partial x^2} + \frac{\partial^2 T}{\partial y^2} + \frac{\partial^2 T}{\partial z^2} \right) + Q_g \quad (4.6)$$

Equations 4.3–4.5 are derived from Newton's second law of motion and are described as momentum conservation equations in the fluid flow. The Gr/Re^2 $(= g\beta(T_{IC} - T_\infty)L_c)$ u^2 term accounts for the effect of mixed convection heat transfer along the airflow direction (x-direction).

The IC chips are supplied with the constant heat flux values as mentioned in Table 3.1; therefore, the energy equation is mentioned only for the IC chips. The 3D, steady-state energy conservation equation is expressed as given in Equation 4.6.

4.3.2 Boundary Conditions

The boundary condition for the problem domain, as shown in Figure 4.3 under the mixed convection heat transfer mode, is given as follows:

Inlet boundary condition:

At X = 0 (inlet to the board), T = T$_\infty$ = 25°C = 298 K, u = 31 m/s, v = w = 0

Outlet boundary condition:

At X = L (exit from the board), P = P$_\infty$

Lateral walls of the domain are assumed to be adiabatic:
$$\frac{\partial T}{\partial x} = \frac{\partial T}{\partial y} = \frac{\partial T}{\partial z} = 0$$

The specifications of the computational model (substrate board with IC chips) used for the present analysis are already mentioned in Table 3.1. Mixed convection is generally preferred in the fan cooling systems, and occurs when the inertia and buoyancy forces are balanced with each other. Mixed convection is characterized by the Richardson number, Ri, which is expressed as Gr/Re^2. Perfect mixed convection is a balance between buoyancy force and inertia force leading to the Richardson number of the order 1 (Ri ≡ 1). The SMPS board is always accompanied by a fan which has to run at the optimal speed for the cooling of different components mounted on it. Considering the realistic approach, in the present study, the mixed convection velocity is calculated as 31 m/s (calculations for the same are given in Appendix C). For this velocity, the Reynolds number is calculated as 33,000 and, hence, the flow along the IC chips is assumed to be laminar. (The problem is modelled as flow over the flat plate for which the critical Reynolds number value, Re_{cr} is 5×10^5).

4.3.3 Grid Independence Study

The grid independence study is necessary to determine the optimal grid size at which the numerical simulations are to be carried out. The pre-processor, processor, and post-processing sections are integrated into one module using the ANSYS Icepak. The processor uses the FLUENT solver based on the control volume technique. All the components in the test section are of standard geometries (rectangular). Therefore, the components have meshed in the pre-processor with HEXA-dominant mesh which gives better results. The computational domain under the mixed convection heat transfer consists of 1,320,644 nodes and 1,208,484 HEXA elements. It is seen that increasing the mesh elements beyond these values doesn't affect much on the maximum temperature difference between the higher node elements (the difference is only 0.67°C). The grid size (nodes and Hexas) are selected based on the convergence criteria for the mass, momentum (in x, y, and z-direction), and energy conservation equations. The convergence criteria are set at $1e^{-8}$ for the energy equation and $1e^{-3}$ for both the mass and momentum equations, respectively. The safe temperature limit (less than 100°C) for the IC chips is also taken into consideration during the grid independence study.

Hence, to save computational time and to have better results, the simulations are carried out for all the 47 selected configurations with a mesh profile having 1,320,644 nodes, 1,208,484 HEXA elements, and 131,504 quads. The mesh profile for a particular configuration $\lambda = 1.78906$ is shown in Figure 4.4. The grid independence study for the same configuration is given in Table 4.2 and its corresponding plot is shown in Figure 4.5.

4.4 RESULTS AND DISCUSSION

The seven asymmetric rectangular IC chips can be arranged in 863,040 different ways on the SMPS board which are designated by a unique value of λ. In the present case, λ varies from 0.25103 to 1.87025. For the

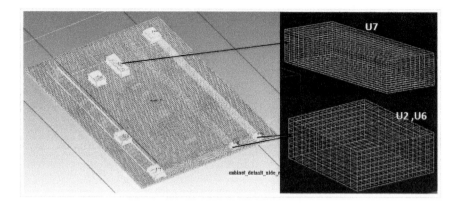

FIGURE 4.4   Mesh profile for the computational domain of $\lambda = 1.78906$.

TABLE 4.2   Grid Independence Study for $\lambda = 1.78906$

S. No.	Nodes	Hexa Elements	T_{max}, °C
1.	640842	686178	70.14
2.	993026	1052400	74.36
3.	1188430	1090440	76.68
4.	**1320644**	**1208484**	**77.15**
5.	1474756	1550228	77.82
6.	1655056	1515744	77.85

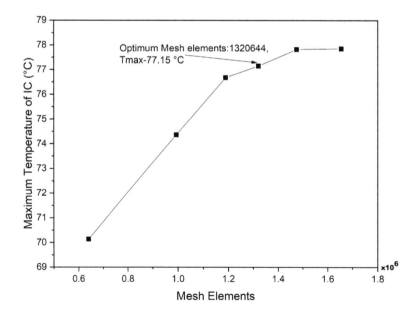

FIGURE 4.5   Grid independence study for $\lambda = 1.78906$.

present study, 47 different configurations are randomly selected for the analysis. The criterion for the selection of these configurations is already explained under Section 4.2. The λ depends strongly on the positions of the IC chips on the substrate board and their size. The goal is to predict the optimal configuration for the arrangement of these seven IC chips. The optimal configuration leads to the minimum of the maximum temperature excess $(T_{max} - T_{\infty})$ among all the possible arrangements (863,040 number of ways) of the IC chips on the SMPS board. The temperature excess is considered, as the IC chips are subjected to different ambient conditions during their working cycle.

4.4.1 Maximum Temperature Excess Variation of Different Configurations with λ

A plot for the maximum temperature excess of the configuration (arrangement of the seven IC chips) with their corresponding λ (as shown in Figure 4.6) suggests that, with the increase in λ, there is a decrease in the maximum temperature excess of the configuration. This is mainly due to

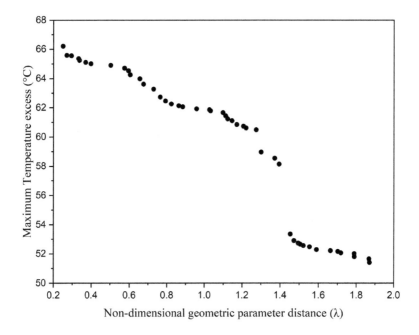

FIGURE 4.6 Variation of maximum temperature excess of different configurations with λ.

an increase in λ, the spacing between the IC chips of a particular configuration increases; thus, the thermal interaction between these IC chips has vanished, and, ultimately, the maximum temperature of the configuration has decreased. It is also observed from Figure 4.6 that the maximum temperature excess among all the configurations is only 67°C, which leads to the maximum IC chip temperature as 92°C (considering the ambient temperature, T_∞ as 25°C). Thus, the IC chips are under the safe operating temperature limit (less than 100°C).

4.4.2 Temperature Variation for the IC Chips of the Lower (λ = 0.25103) and the Upper Extreme (λ = 1.87025) Configurations

The temperature variation for the IC chips of the two extreme (lowest and highest) configurations, λ = 0.2510 (34-33-55-24-35-25-23) and λ = 1.87025 (72-71-55-24-75-11-23) is shown in Figure 4.7. It is observed that the lowest extreme configuration, λ = 0.2510 leads to the maximum temperature

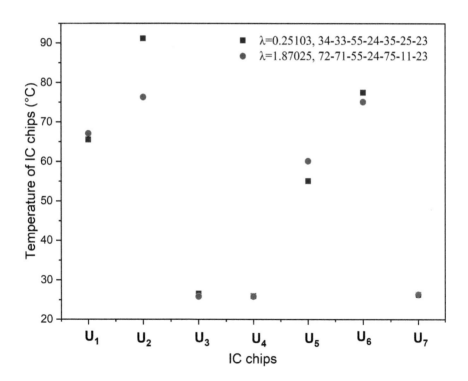

FIGURE 4.7 Temperature variation for the IC chips of two configurations; λ = 0.25103 and λ = 1.87025.

among the IC chips and the maximum temperature is minimum for the upper extreme configuration, $\lambda = 1.87025$. The temperature difference between these two configurations is in the ranges of 4–8% only.

For both configurations, the IC chip U_2 attains the maximum temperature due to its smaller size and high heat dissipation rate. This results in more heat accumulation inside the chip leading to an increase in its temperature. Again, the low-powered IC chips U_3 (55), U_4 (24), and U_7 (27) have a very low temperature (close to ambient) irrespective of the configuration, as their heat dissipation rates are very low. This confirms the importance of the IC chip's size and their input heat flux value for their cooling.

4.4.3 Empirical Correlation

To generalize the whole discussion, the maximum temperature excess $(T_{max} - T_{\infty})$ of the configuration (obtained from the numerical simulation) is converted into the non-dimensional maximum temperature excess $\theta = \dfrac{T_{max} - T_{\infty}}{\Delta T_{ref}}$, $\Delta T_{ref} = \dfrac{qL_c}{k_f}$. A correlation is then put forth for the θ in terms of λ, and is given in Equation 4.7.

$$\theta = 0.008(1+\lambda)^{-0.327} \tag{4.7}$$

Equation 4.7 is based on 47 data points. It has a regression coefficient (R^2) of 0.81, a root-mean-square (RMS) error of $\pm3.5\%$ on the estimate, and is valid for $0.25103 \le \lambda \le 1.87025$, $0.0067668 \le \theta \le 0.0051602$. To check the accuracy of this correlation, a parity plot is plotted between the theta simulation (θ_{sim}), and theta correlation (θ_{corr}), and is shown in Figure 4.8. It is seen that both the data agree well with an error band of $\pm10\%$.

Using Equation 4.7, the θ_{corr} values are calculated for the remaining 862,993 (863,040–47) configurations using their λ values. The variations of θ_{corr} for all the 863,040 configurations are then plotted with their corresponding λ and are shown in Figure 4.9. A similar trend is followed (decrease in the maximum temperature with the increase in λ) for all the configurations, as is clear from Figure 4.9. The temperature variation between the lowest and highest extreme configurations is only 8% which is within the permissible range, as explained under Section 4.2. Hence, it is understood that Equation 4.7 can capture the whole physics embedded in the problem under consideration.

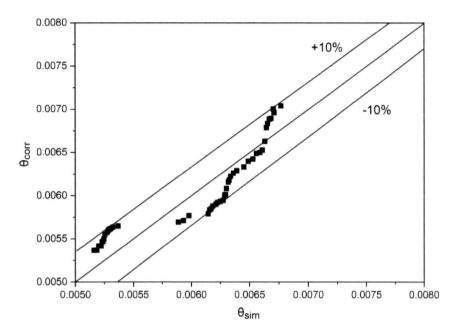

FIGURE 4.8   Error plot between θ_{sim} and θ_{corr}.

However, Equation 4.7 is based on 47 data points only which are used to extrapolate the θ values for the remaining 862,993 configurations to predict the global optimal among them. This is an unrealistic approach; hence, to minimize the error between the temperatures obtained by the

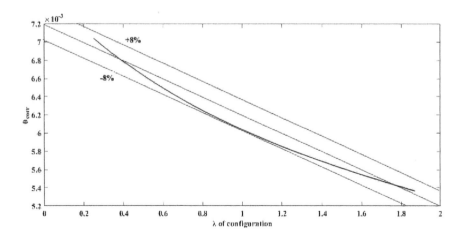

FIGURE 4.9   Variation of θ for all the 863,040 configurations with λ: (Validation of Equation 4.7).

prediction (correlation) and the numerical simulation, a robust technique is needed, which is explained under Section 4.5.

4.5 HYBRID OPTIMIZATION STRATEGY

A hybrid optimization strategy is implemented to consider all the higher-order effects embedded in the problem under consideration. This uses the numerical data-driven combined artificial neural network (ANN) and genetic algorithm (GA) strategy.

4.5.1 Artificial Neural Network

ANN has become increasingly prevalent in solving engineering problems particularly in convective fluid flow and heat transfer. The technique is very quick and has the advantage of reducing the computational time considerably. The ANN works similarly to the functioning of the human brain. A neural network consists of a different number of neurons forming a hidden layer. These neurons are connected with input and output layers forming a neuron network. Each network connection has a weight and each neuron is associated with a bias value. The output is predicted using the input vectors and varying the weight in the network connection which is called network training. The training is carried out to reduce the error between the actual and network output.

Initially, a forward model is developed using numerical simulation data. The neural network is trained for all the 863,040 possible configurations with λ as the input and θ_{corr} as the output (obtained using the correlation, Equation 4.7). The training is carried out using a feed-forward backpropagation algorithm with one hidden layer. The transfer function used is Tan-sigmoid and Levenberg-Marquardt algorithm is used for updating the weights and bias values of the network. The training is carried out using the neural network toolbox in MATLAB 2015a. The training of the network is carried out using 75% data, and the remaining 25% data is used for testing the network. A parity plot between the ANN outputs and the correlated output values (using Equation 4.7) with the 25% testing data is shown in Figure 4.10. It confirms a good agreement of both the data with an error band of 10% and thus confirms the accuracy of the network.

To check the accuracy of the network, a neuron independence study is carried out by varying the number of neurons in the hidden layer from 2 to 20. The optimal network is determined with the least value of mean square error (MSE) and mean relative error (MRE), and the highest value

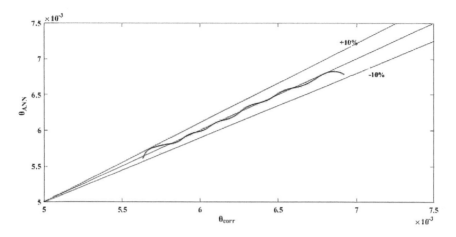

FIGURE 4.10 Parity plot between θ_{corr} and θ_{ANN}.

of regression coefficient (R^2). The calculations for MSE, MRE, and R^2 are given in Equations 4.8–4.10 and the values are given in Table 4.3.

$$MSE = \frac{1}{n}\sum_{i=1}^{n}\left(T_{corr} - T_{ANN}\right)^2 \tag{4.8}$$

$$MRE = \frac{1}{n}\sum_{i=1}^{n}\frac{\left|T_{corr} - T_{ANN}\right|}{T_{corr}} \tag{4.9}$$

$$R^2 = 1 - \left[\frac{\displaystyle\sum_{i=1}^{n}\left(T_{corr} - T_{ANN}\right)^2}{\displaystyle\sum_{i=1}^{n}\left(T_{corr}\right)^2}\right] \tag{4.10}$$

TABLE 4.3   Results of the Neuron Independence Study

No. of Neurons	MSE	MRE	R^2
2	7.94×10^{-8}	0.0371691	0.78
5	2.27×10^{-9}	0.0054355	0.986
7	2.97×10^{-9}	0.0046627	0.981
10	**1.77×10^{-9}**	**0.0028512**	**0.988**
15	4.15×10^{-9}	0.0056512	0.975
20	5.27×10^{-9}	0.0065286	0.968

From Table 4.3, it is confirmed network with 10 neurons (provided in Bold in Table 4.3) has the least value of MSE and MRE and the highest value of R^2. Hence, this is the optimal network to power the GA. The ANN outputs are then deployed to the GA to determine the globally optimal configuration for the arrangement of seven asymmetric IC chips.

4.5.2 Genetic Algorithm

The role of optimization is critical for design engineers to deal with multi-variable problem-solving techniques. Several optimization techniques like the analytical approach, downhill simplex method, gradient descent, and many more are available. The disadvantage of these algorithms is of getting "stuck" in the local minimum. To overcome this, evolutionary algorithms like GAs, simulated annealing, and ant colony optimization techniques are proposed recently.

The GA (Baby and Balaji (2013)) is a search-based technique that works on the principle of genetics and natural selection. It rapidly searches for the most possible solutions to a given problem. Hence, this method is handy to reach the global optimum by escaping from the local optima. Due to these merits, GA is employed for the present study. As the solution space for the present case is vast, several local optima may be possible, and, hence, GA can be very robust to identify the global optimum very accurately.

GA deals with an initially randomly generated population of possible solutions. The objective function represents the fitness value of each individual set. For crossover, the improved fitter individuals are identified, and the new generation consists of some new and best individuals. GA works with a population or a set of candidate solutions and tries to improve the solution quality with successive iterations. The population variables are coded as binary strings in GA and the population size is specified depending on the accuracy level. The fitness (objective) function is then calculated by substituting the variables. The ANN outputs are used to evaluate the fitness function in the present study. Different operations like reproduction, crossover, and mutation are carried out to create new populations. Reproduction leads to the copying of the individual strings according to their objective function values. When one crossover point is selected, the binary string from the beginning of the chromosome to the crossover point is copied from one parent and the rest are copied from the second parent, and this is known as single-point crossover. Two points uniform, and arithmetic, are the other types of cross-over strategies used for the GA.

The mutation is considered to neutralize the undesired cross-over effects by changing the bits randomly or stochastically. The probability of cross-over fraction is kept as 0.8 for the present study. The GA scheme terminates either after reaching the tolerance limit or reaching the specified number of iterations.

In the present study, the ANN outputs are used to evaluate the fitness function (objective function) for the problem under consideration. The objective of the GA is to minimize the non-dimensional maximum temperature excess, θ. Thus, the objective function is expressed as given in Equation 4.11.

$$\text{Minimize } \theta \tag{4.11}$$

$$\text{Subjected to } 0.25103 \leq \lambda \leq 1.87025 \text{ and}$$

$$0.0581 \text{ W/cm}^2 \leq q \leq 5.46 \text{ W/cm}^2$$

In the present case, the θ is provided by the trained ANN. GA is executed until the convergence criteria have reached, i.e. to determine the fitness value of the population which is the minimum of the non-dimensional maximum temperature excess among all the configurations. For the present case, the population size is considered as 50, and the crossover fraction is 0.8.

4.5.3 Combination of Artificial Neural Network and Genetic Algorithm

ANN is then combined with GA which helps reduce the number of simulations and thus reduces the computational time. Therefore, the GA can optimize quickly and determine the globally optimal configuration of the IC chips. The flow chart shown in Figure 4.11 illustrates the methodology to determine the global optimal position of the seven asymmetric IC chips arranged at different positions on the SMPS board.

The optimal ANN network obtained in Section 4.5.1 is deployed to GA. The numerical simulations are then carried out until convergence criteria have been reached. The GA gives the optimal solution after 51 generations. The fitness value, θ obtained from the GA is 0.00512087, as shown in Figure 4.12. The corresponding λ value for this is 1.58830. The effect of generations on the fitness value of the optimal configuration is also studied and it is seen that increasing the number of generations has a negligible effect on the maximum temperature of the configuration.

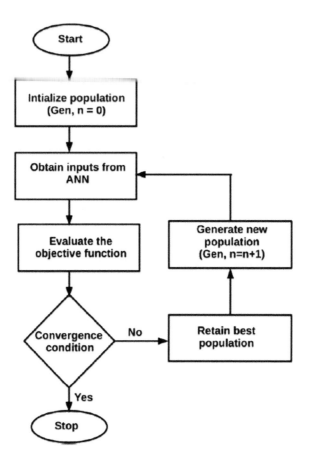

FIGURE 4.11   Flow chart to determine the optimal configuration of the IC chips using GA.

The θ value of the configuration is then converted into the maximum temperature of 75.03°C. The optimal configuration (generated by GA) is shown in Figure 4.13, where the positions of the IC chips are 15-71-55-24-75-11-23 (λ=1.58830) for U_1, U_2, U_3, U_4, U_5, U_6, and U_7, respectively. This method of combining the ANN with GA is known as a hybrid optimization technique. The numerical simulation is further carried out for the optimal configuration obtained from the GA to determine the maximum configuration temperature, and the value is found to be 78.44°C. It is seen that both the simulation and optimization results agree well with each other with an error band of 4.34%.

It is clear from Figure 4.13 that the IC chips U_3, U_4, and U_7 have very low temperatures (close to ambient) due to their low heat dissipation rates,

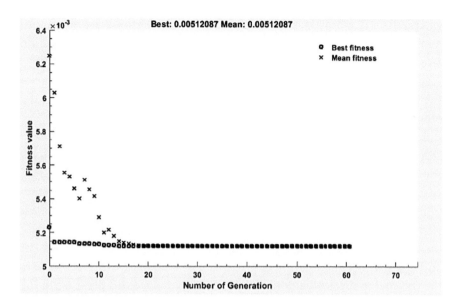

FIGURE 4.12   Fitness value of the optimal configuration using GA.

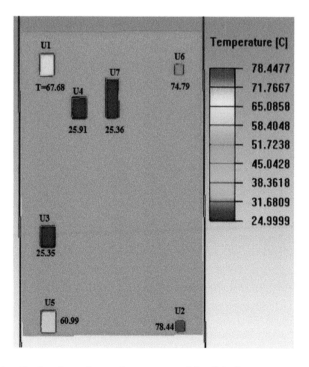

FIGURE 4.13   Optimal configuration generated by GA, $\lambda = 1.58830$.

whereas the IC chips U_2 and U_6 are identical, and attain the maximum temperature in the configuration. Due to the maximum heat dissipation rate, the IC chip U_2 attains the maximum temperature. Hence, this chip is positioned near the domain outlet which is located at the extreme edge of the substrate board. This helps reduce the thermal interaction among the other IC chips and allows the chips to cool fast.

Figure 4.14 (top) signifies the fluid passing at the lower surface of the IC chips for which there is no thermal interaction and Figure 4.14 (bottom) shows the fluid passing through the top (upper) surface of the IC chips, which shows their heat dissipation rate along with their thermal interaction. This is visualized from the contour plots of Figure 4.14. The hottest IC chip U_2 is placed near the outlet port and the IC chip U_6 is placed near the inlet port of the domain. The two hottest and smallest IC chips (U_2 and U_6) are positioned at the opposite end of the board which minimizes the thermal interaction among the other chips and led to a decrease in the configuration temperature.

4.6 CONCLUSIONS

3D steady-state numerical simulations are carried out on seven asymmetric rectangular IC chips mounted at different positions on an SMPS board cooled using air under the mixed convection heat transfer mode. The goal is to determine its optimal configuration. The following conclusions are obtained from the study:

1. Temperature of the IC chips is a strong function of their size and position on the substrate board.

2. The size of the IC chips and their positions on the board are characterized by the non-dimensional position parameter, λ. The highest λ signifies that the maximum temperature of the configuration is minimum, thus leading to the optimal.

3. The smallest size IC chip, U_2 with a high heat dissipation rate attains the maximum temperature in the configuration. However, the temperature variation for the low-powered IC chips U_3, U_4, and U_7 are very small (close to ambient).

4. A correlation between the θ and λ can predict the entire physics of the problem under consideration.

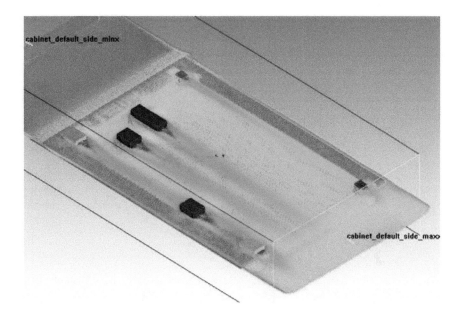

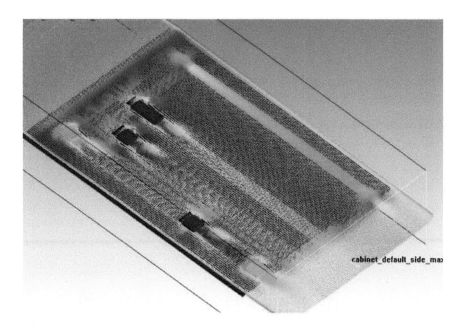

FIGURE 4.14 Velocity profile for the optimal configuration obtained using GA.

5. The numerical data-driven combined ANN-GA-based strategy predicts the optimal configuration of the IC chips more accurately as compared to the conventional one.

6. The optimal configuration obtained from the ANN-GA-based strategy confirms that the IC chips dissipating more heat must be placed near the exit port of the domain so there is negligible thermal interaction among the other IC chips.

Hybrid Optimization Strategy to Study the Substrate Board Orientation Effect for the Cooling of the IC Chips under Forced Convection

5.1 INTRODUCTION

The chapter deals with the results of the experimental and numerical investigations carried out for seven asymmetric protruding IC chips mounted at different positions on the switched-mode power supply (SMPS) board cooled under the laminar forced convection heat transfer. The SMPS board is oriented at different angles, i.e. at 0° (horizontal), 30°, 60°, and 90° (vertical), to study their effect on the cooling of the IC chips. The goal is to determine the optimum configuration for the arrangement of the seven IC chips on the board leading to minimize their maximum temperature. The optimal configuration is determined using a hybrid optimization strategy [combined artificial neural network (ANN) and genetic algorithm (GA)]. Both the experimental and numerical results are then compared with each

DOI: 10.1201/9781003188506-5

other. Different correlations are proposed to predict the temperature of the IC chips for different substrate board orientations. For a better understanding, the chapter is classified into the following sections:

- Use of the non-dimensional geometric distance parameter (λ)

- Experimental methodology

- Effect of substrate board orientation on the temperature of the IC chips

- Optimal arrangement of the seven IC chips using hybrid optimization technique

- Numerical framework

5.2 DIFFERENT IC CHIPS COMBINATIONS CONSIDERED FOR EXPERIMENTATION

Steady-state experiments are then conducted for 32 different configurations of the IC chips, as mentioned under Table 5.1. The seven asymmetric rectangular IC chips are supplied with non-uniform heat fluxes as mentioned in Table 3.4. The detailed experimental procedure

TABLE 5.1   Different IC Chip Configurations Selected for the Experiments

λ	U_1-U_2-U_3-U_4-U_5-U_6-U_7	λ	U_1-U_2-U_3-U_4-U_5-U_6-U_7
0.25103	34-33-55-24-35-25-23	1.06064	42-64-55-24-75-15-23
0.27076	34-35-55-24-45-25-23	1.11461	71-52-55-24-15-31-23
0.29491	14-34-55-24-15-25-23	1.16858	71-62-55-24-35-51-23
0.35898	12-33-55-24-35-25-23	1.22256	64-72-55-24-54-11-23
0.41295	32-45-55-24-53-34-23	1.27653	71-12-55-24-15-62-23
0.46693	33-15-55-24-25-11-23	1.3305	75-61-55-24-14-65-23
0.5209	43-54-55-24-31-45-23	1.38448	63-75-55-24-72-73-23
0.57487	41-52-55-24-43-54-23	1.43842	25-75-55-24-72-61-23
0.62885	43-64-55-24-63-34-23	1.49245	73-74-55-24-14-75-23
0.68282	65-32-55-24-12-44-23	1.54642	25-73-55-24-74-71-23
0.7368	63-42-55-24-32-11-23	1.60047	74-73-55-24-72-71-23
0.79077	34-63-55-24-41-64-23	1.65489	71-74-55-24-75-21-23
0.84474	34-64-55-24-73-42-23	1.70834	75-71-55-24-13-72-23
0.89872	53-64-55-24-62-14-23	1.765	75-12-55-24-71-72-23
0.95269	15-52-55-24-44-73-23	1.81349	71-75-55-24-72-11-23
1.00666	11-71-55-24-35-14-23	1.87025	72-71-55-24-75-11-23

for the laminar forced convection heat transfer mode is explained under Section 3.5.1. The experiments are repeated for two different air velocities of 4.5 m/s and 8 m/s, respectively, and different substrate board orientations (δ), i.e. 0°, 30°, 60°, and 90°, as shown in Figure 3.17. The calculations for different parameters of the experiment are carried out as explained under Section 3.6.1 using Equations 3.2–3.9. The schematic diagram of different configurations and the photographic view of two configurations used for the experimental analysis are shown in Figures 5.1 and 5.2, respectively (Mathew and Hotta (2020)). Three correlations are proposed, the first one is for the non-dimensional maximum temperature excess (θ) of the configuration in terms of their λ value; the second one is for the non-dimensional temperature excess θ_i of each chip in terms of their non-dimensional position ($Z = X_i/Y_i$) on the substrate board, non-dimensional substrate board orientation (φ = actual angle/90°) and their sizes ($S = t_c/l_c$), and the third one is for the Nusselt number (Nu) of the IC chips in terms of the fluid Reynolds number (Re) and the IC chip's size (S).

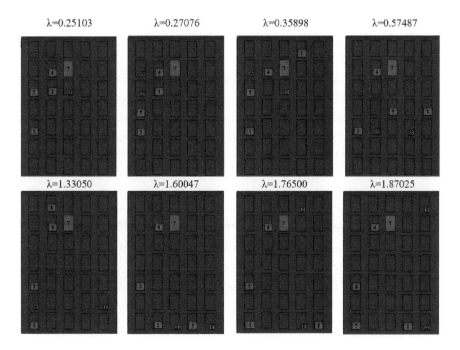

FIGURE 5.1 Schematic diagram of different configurations used for the experimental analysis.

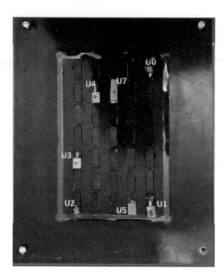 

FIGURE 5.2   Photographic view of two configurations used for the experimental analysis.

5.3 RESULTS AND DISCUSSION

5.3.1 Temperature Variation of the IC Chips for Different Substrate Board Orientations

The temperature variation of the IC chips for different substrate board orientations for a particular configuration, $\lambda = 0.25103$ (lower extreme) under a velocity of 4.5 m/s is shown in Figure 5.3 (top). It is seen that the IC chip U_2 attains the maximum temperature in the configuration for all four substrate board orientations. This is because U_2 has the smallest size and its heat dissipation rate is also high. Hence, this chip accumulates more heat, and its temperature rises. Again, the temperature of this chip for the 30° board orientation is very less as compared to the other orientations. The substrate board is fixed at the center of the test section of the wind tunnel so that the free stream velocity becomes maximum while entering from the inlet port. When the substrate board is inclined, the velocity boundary layer is developed from the leading edge of the IC chip. The seven IC chips are having different thermal boundary layers due to the change in their length and heat flux values. The thermal boundary layer decreases with the increase in the inclination of the substrate board which leads to a decrease in the temperature of the IC chips. Therefore, the maximum temperature of the IC chips is found to be minimum for the 30° orientation in which

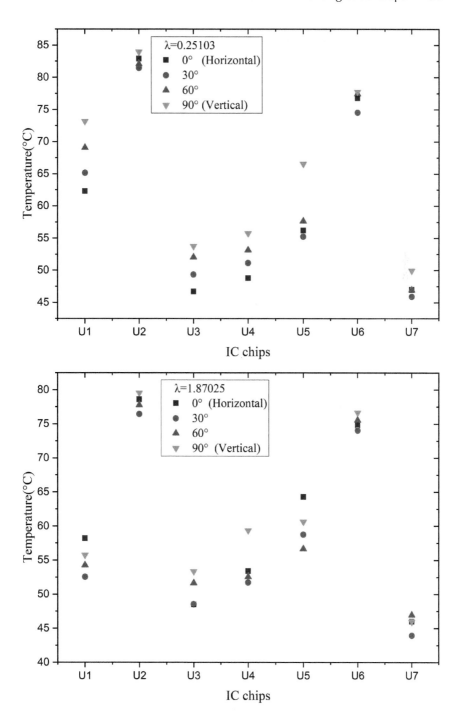

FIGURE 5.3 Temperature of the IC chips for different substrate board orientations.

the velocity boundary layer is not disturbed and there is no formation of the vortices. When the orientation of the substrate board is 60° and 90°, the thickness of the hydrodynamic boundary layer increases and the formation of vortices takes place which disturbs the thermal boundary layer due to which there is greater thermal interaction between the IC chips and the maximum temperature of the configuration rises. The temperature of the IC chips concerning the board orientation follows 30° < 0° < 60° < 90°.

There are some anomalies to this trend for the largest size IC chips, as the surface area of these chips is predominant as compared to the substrate board orientation effect. Similar trends are observed for the upper extreme ($\lambda = 1.87025$) configuration, as shown in Figure 5.3 (bottom). There is a temperature drop of about 5°C between the lower and upper extreme configurations leading to cooling the IC chips faster.

Figure 5.4 shows the maximum temperature variation of various configurations (arrangement of seven IC chips) for different substrate board orientations. It is seen that the maximum temperature of the configuration is lowest for 30° leading to faster cooling of the IC chips. As the air hits

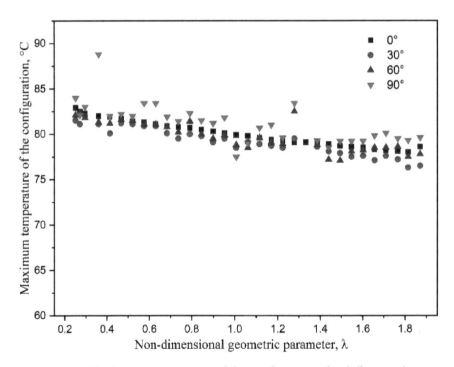

FIGURE 5.4 Maximum temperature of the configuration for different substrate board orientations.

perpendicular to the board for 90°, the thermal and the velocity boundary layer is disturbed, which causes a distorted flow. Thus, the thermal interaction between the chips increases, which leads to an increase in the maximum temperature of the configuration. There is a temperature drop of 2–3°C between these two configurations. The maximum temperature is dropped significantly by increasing the λ value of the configuration, as by doing so, the geometric spacing of the IC chips increases. This lowers down the thermal interaction among the IC chips, and, hence, the configuration temperature drops. There is a temperature drop of about 5°C between the lower and upper extreme configurations leading to better thermal management. The main objective of the study is to minimize the maximum temperature of the IC chips for their optimal positioning on the board with optimal orientation.

5.3.2 Temperature Variation of IC Chips for Different Air Velocities

The temperature variation of the IC chips for their lower and upper extreme configurations under two different air velocities (4.5 m/s and 8 m/s) for the horizontal substrate board is shown in Figure 5.5. It is seen that the temperature of the IC chips is lowered at the higher velocities for all the configurations, as at higher velocities, the rate of heat dissipation from the chips is more. There is a temperature drop of 1.42–17.33% of the IC chips for the 8 m/s velocities as compared to the 4.5 m/s velocities. All the remaining 30 configurations also follow a similar trend. Again, the IC chip U_2 attains the maximum temperature in the configuration due to its maximum heat dissipation rate.

5.3.3 Maximum Temperature Variation of the Configurations for Different Substrate Board Orientations

Figure 5.6 shows the variation of maximum temperature for different configurations under different substrate board orientations. It clearly shows that the maximum temperature of the configurations reduces with the increase in their λ value.

For all the cases, the IC chip U_2 attains the maximum temperature due to its high heat dissipation rate. The thermal boundary layer thickness is minimum for the 30° orientation and it is disturbed due to the increase in velocity boundary layer and formation of the vortices for the 60° and 90° orientations. The other configurations follow the relation 30° < 0° < 60° < 90° for their maximum temperature. By increasing the velocity from 4.5 m/s to 8 m/s, the maximum temperature of the configuration drops by 1.42–17.33%.

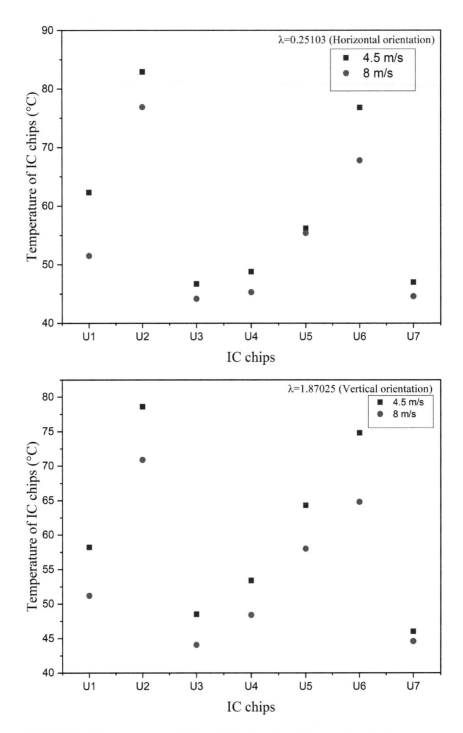

FIGURE 5.5   Temperature variation of IC chips for different air velocities.

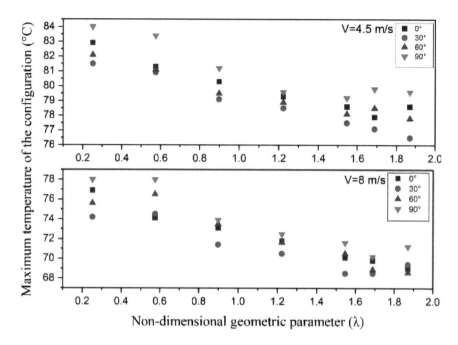

FIGURE 5.6 Configuration maximum temperature variation for different substrate board orientations.

5.3.4 Variation of Maximum Heat Transfer Coefficient of the Configurations for Different Substrate Board Orientations

The heat balance showing the contribution of the convection rate is mentioned under Appendix D. The maximum convective heat transfer coefficient variation of the configurations for different substrate board orientations is shown in Figure 5.7. By increasing the λ value, the maximum convective heat transfer coefficient of the configuration increases, which signifies better cooling of the IC chips and ultimately leads to their better thermal performance. Similar trends are observed for all the substrate board orientations. Among the different orientations, the 30° board has shown better cooling.

5.4 EMPIRICAL CORRELATION

The main goal of the present work is to minimize the maximum temperature of IC chips in determining their optimal configuration and optimal size. Most of the portable electronic gadgets are mounted on a horizontal substrate board. Hence, the horizontal substrate board is preferred to determine the optimal configuration.

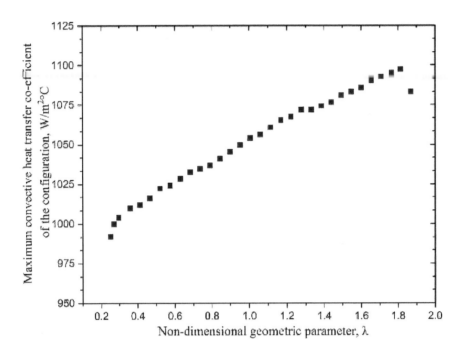

FIGURE 5.7 Variation of maximum heat transfer co-efficient of the configurations for different substrate board orientations.

5.4.1 Correlation for θ in Terms of λ

A correlation is put forth for the non-dimensional maximum temperature excess (θ) of the IC chips in terms of their non-dimensional geometric parameter distance (λ). The correlation is given in Equation 5.1.

$$\theta = 0.005(1+\lambda)^{-0.118} \tag{5.1}$$

Equation 5.1 is based on 32 configurations of λ, has a coefficient of regression of 0.97 with a root mean square (RMS) error of 0.012%, and is valid for the following parameter ranges: $0.25103 \leq \lambda \leq 1.87025$ and $0.005015 \leq \theta \leq 0.004568$. A parity plot between the θ_{exp} and θ_{corr} (obtained using Equation 5.1) shown in Figure 5.8 suggests a strong agreement between these two values with an error band of 12%.

Now, Equation 5.1 is used to extrapolate the temperature data for all the remaining configurations, i.e. 863,008 (863,040–32). The variation of θ_{corr} for all the 863,040 configurations with their λ values is shown in Figure 5.9 which depicts that θ_{corr} decreases with increase in λ of the configuration. The maximum temperature is minimum for the upper extreme value, $\lambda = 1.870255$.

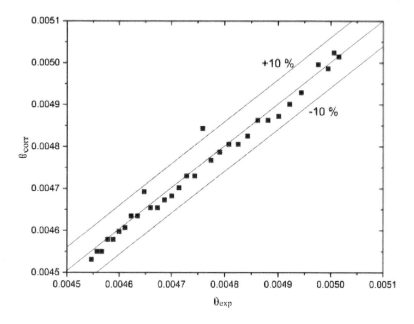

FIGURE 5.8 Parity plot between θ_{exp} and θ_{corr}.

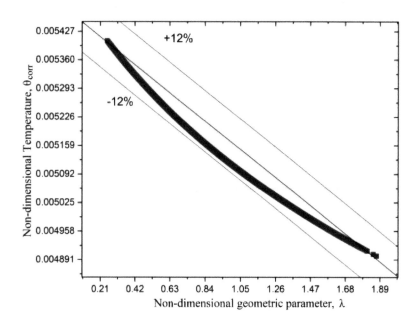

FIGURE 5.9 Validation of θ (obtained from Equation 5.1) with λ.

The objective is to determine the optimal arrangement of the IC chips among the 863,040 configurations. However, Equation 5.1 is based on 32 data points only. Hence, to minimize the error for the θ extrapolated values of all the 863,040 configurations, a robust optimization strategy (combined ANN and GA) is adopted which is discussed under Section 5.5.

5.4.2 Correlation for θ_i in Terms of the IC Chip Positions on the Substrate Board (Z), Non-Dimensional Board Orientation (φ), and IC Chip Sizes (S)

A correlation is put forth for the non-dimensional temperature excess of each chip (θ_i) of the configuration in terms of their position on the substrate board (Z), non-dimensional substrate board orientation (φ), and their sizes (S). The correlation is given in Equation 5.2.

$$\theta_i = 0.017 \left(1+Z\right)^{0.0361} \left(1+\cos\varnothing\right)^{-0.079} S^{1.279} \tag{5.2}$$

Equation 5.2 has a regression coefficient of 0.88, an RMS error of 0.038% and is valid for the following parameter ranges, $0.001458 \le \theta_i \le 0.005413$; $0.105163 \le Z \le 19.54713$; and $0 \le \varphi \le 1$, $0.17455 \le S \le 0.35$.

5.4.3 Correlation for Nusselt Number of the IC Chips in Terms of Fluid Reynolds Number and IC Chip's Size

As discussed in Figure 5.7, convective heat transfer plays a vital role in the cooling of the IC chips. Hence, a correlation is developed for the Nusselt number of the IC chips in terms of the fluid Reynolds number and the IC chip's size. The correlation is given in Equation 5.3.

$$Nu = 0.238\left(1+Re\right)^{0.786} S^{-0.83} \tag{5.3}$$

Equation 5.3 has a regression coefficient of 0.92, an RMS error of 0.37%, and is valid for $1{,}251 \le Re \le 2{,}751$, $0.17455 \le S \le 0.35$, and $137 \le Nu \le 631$. Figure 5.10 shows the Nusselt number variation of the IC chips with the fluid Reynolds number for all the configurations, which shows that the Nusselt number improves with the increase in Reynolds number, and the IC chip U7 has the highest Nusselt number. Equation 5.3 is used to predict the Nusselt number of the IC chips based on their size and fluid Reynolds number. Figure 5.11 shows the parity plot between the Nu_{exp} and Nu_{corr}, and it suggests that both the data agree within 15% error.

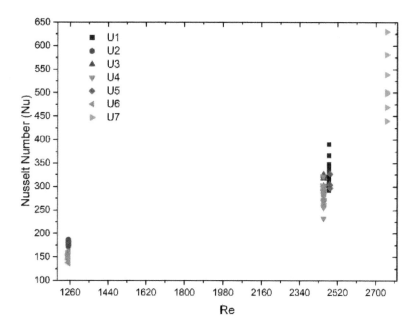

FIGURE 5.10   Nusselt number variation of the IC chips with fluid Reynolds number and the IC chip's size.

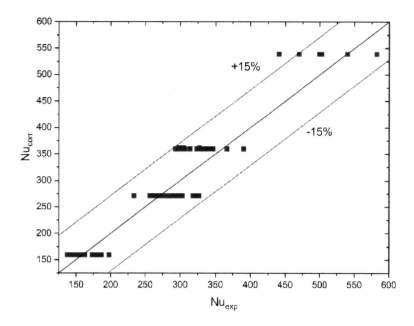

FIGURE 5.11   Parity plot between Nu_{exp} and Nu_{corr}.

5.5 HYBRID OPTIMIZATION STRATEGY TO IDENTIFY THE OPTIMAL BOARD ORIENTATION AND OPTIMAL CONFIGURATION OF THE IC CHIPS

The method of determining the optimal configuration of the IC chips using the combined ANN-GA approach is already detailed under Section 4.5 of Chapter 4.

5.5.1 Artificial Neural Network

For the present study, the artificial neural network is developed using experimental data. The correlation proposed in Equation 5.5 using the experimental data is used to predict the maximum temperature for all the 836,040 configurations. Hence, in the present study, the neural network is trained for all the 836,040 configurations with λ as the input and θ_{corr} as the output to determine the optimal network. The neural network is trained using the MATLAB® 2015a toolbox using 70% training data and 30% testing data. The network is also trained for different hidden layers with the number of neurons in each layer varying from 2 to 14. The performance of the neural network is based on three parameters: mean square error (MSE), mean relative error (MRE), and Regression coefficient (R^2), the expressions for the same are given in Equations 4.8–4.10. It is found that the network with 6 neurons has the lowest value of MSE and MRE and the highest value of R^2; hence, this is the optimal network. The neuron independence study of the network is given under Table 5.2. More details about the artificial neural network are already given under Section 4.5.1 in Chapter 4.

After obtaining the optimal neural network, an error plot is plotted for the temperature data obtained from the ANN (θ_{ANN}) and the correlation

TABLE 5.2   Results of the Neuron Independence Study

No. of Neurons	MSE	MRE	R^2
2	3.68×10^{-9}	0.010568	0.85445
4	2.25×10^{-9}	0.009031	0.972606
6	$\mathbf{2.20 \times 10^{-9}}$	**0.019098**	**0.979944**
8	1.00×10^{-9}	0.008125	0.846711
10	1.56×10^{-9}	0.019262	0.835659
12	4.28×10^{-9}	0.010865	0.757064
14	2.49×10^{-9}	0.007463	0.863114

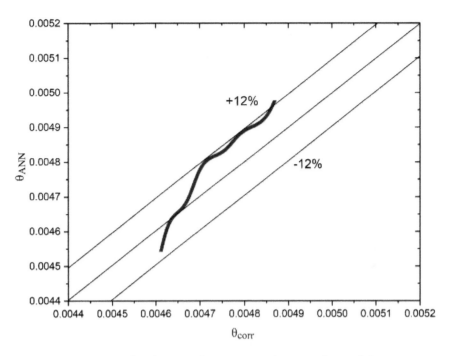

FIGURE 5.12   Parity plot showing the agreement between θ_{corr} and θ_{ANN}.

(θ_{corr}) and is shown in Figure 5.12. Both the data agree well with an error band of 12% and thus confirm the accuracy of the network. The optimal network is then chosen to power the GA. The ANN output is then deployed to the GA to determine the global optimal configuration for the arrangement of seven asymmetric IC chips.

5.5.2 Genetic Algorithm

The details about the execution of the genetic algorithm are already explained under Section 4.5.2. For the present study, the optimal ANN network is employed to GA for the calculation of fitness function (objective function). The objective of GA is to minimize the non-dimensional maximum temperature, θ and is given in Equation 5.4.

$$\textit{minimize } \theta \tag{5.4}$$

Subjected to $0.25103 \leq \lambda \leq 1.87025$ and $1.1743 \text{ W/cm}^2 \leq q \leq 5.46 \text{ W/cm}^2$.

The optimization toolbox in MATLAB 2015a is used to execute the GA until the convergence has reached. This gives the minimum of the non-dimensional maximum temperature excess among the 863,040 configurations. The population size considered for the present study is 51 and the crossover function is 0.8.

5.5.3 Combination of ANN and GA

The optimal ANN network obtained in Section 5.5.1 is then deployed to GA. The GA is executed until the convergence criteria have reached and given the optimal solution after 51 generations. The fitness value, θ obtained from the GA is 0.00444798, as shown in Figure 5.13. The corresponding λ value for this is 1.68806. The effect of generations on the fitness value of the optimal configuration is also studied and it is seen that increasing the number of generations has a negligible effect on the maximum temperature of the IC chips. The flowchart for this optimization methodology is already given in Figure 4.11.

The non-dimensional maximum temperature excess, θ of the configuration is then converted into the maximum temperature using the expression given under the nomenclature section. The maximum temperature is

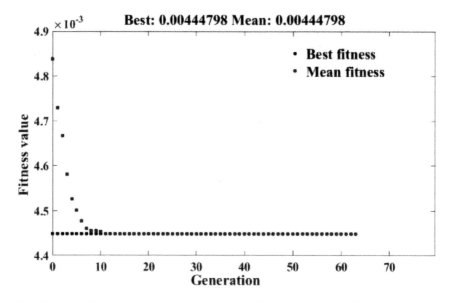

FIGURE 5.13   Fitness value of the optimal configuration using GA.

found to be 76.92°C and the optimal configuration ($\lambda = 1.68806$) is 71-75-55-24-61-11-23 for the IC chips U_1, U_2, U_3, U_4, U_5, U_6, and U_7, respectively. This method of combining the ANN and GA is known as a hybrid optimization strategy.

To validate the GA results, steady-state experiments are further conducted for the optimum configuration obtained from GA ($\lambda = 1.68806$) to evaluate the maximum temperature of the configuration. The T_{max} obtained from the experimental analysis is 77.9°C, which agrees with the GA results with an error of 2.23%. The results suggest that the IC chips with maximum temperature and maximum heat flux must be placed at the extreme edge of the substrate board near to the outlet port, so that there will be negligible thermal interaction among the other IC chips and the configuration temperatures may not be affected. The temperature distribution of the seven IC chips is shown in Figure 5.14. This clearly shows that the IC chips are operated below the critical temperature (less than 100°C) and hence, their reliability and performance are enhanced.

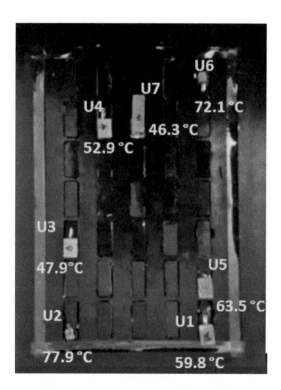

FIGURE 5.14   Optimal configuration generated by GA, $\lambda = 1.68806$.

5.6 NUMERICAL INVESTIGATION FOR THE COOLING OF THE SEVEN ASYMMETRIC IC CHIPS UNDER THE LAMINAR FORCED CONVECTION

Steady-state numerical simulations are carried out on all the configurations for the arrangement of seven asymmetric IC chips as mentioned in Table 5.1 under the laminar forced convection heat transfer mode. The test section, substrate board, and IC chips are modeled as per the dimensions used for the experimental analysis. The materials taken for the IC chips are aluminium and the substrate board is Bakelite.

5.6.1 Computational Model with Governing Equations

The computational model used for the present numerical simulations (carried out using ANSYS Fluent V16.0) is shown in Figure 5.15.

Steady-state governing equations for the continuity, momentum, and energy are solved using the computational fluid dynamics (CFD) solver fluent (ANSYS FLUENT V16.0) and are mentioned in Equations 5.5–5.9. The 3D steady-state mass conservation equation (the continuity equation) is expressed as given in Equation 5.5.

$$\frac{\partial u}{\partial x} + \frac{\partial v}{\partial y} + \frac{\partial w}{\partial z} = 0 \qquad (5.5)$$

Although the air is compressible, its density variation is negligibly small and is assumed to be constant for the temperature range considered in the study.

The 3D steady-state momentum equation in an inertial reference frame is expressed as given in Equations 5.6–5.8. These are derived from Newton's

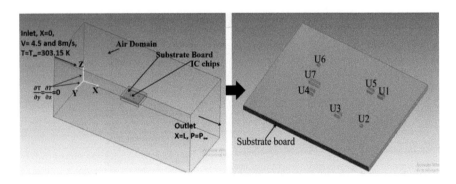

FIGURE 5.15   Computational model used for the present numerical analysis.

second law of motion and are described as momentum conservation equations for the fluid flow.

$$u\frac{\partial u}{\partial x}+v\frac{\partial u}{\partial y}+w\frac{\partial u}{\partial z}=-\frac{\partial u}{\partial x}+\mu\left(\frac{\partial^2 u}{\partial x^2}+\frac{\partial^2 u}{\partial y^2}+\frac{\partial^2 u}{\partial z^2}\right) \tag{5.6}$$

$$u\frac{\partial v}{\partial x}+v\frac{\partial v}{\partial y}+w\frac{\partial v}{\partial z}=-\frac{\partial v}{\partial x}+\mu\left(\frac{\partial^2 v}{\partial x^2}+\frac{\partial^2 v}{\partial y^2}+\frac{\partial^2 v}{\partial z^2}\right) \tag{5.7}$$

$$u\frac{\partial w}{\partial x}+v\frac{\partial w}{\partial y}+w\frac{\partial w}{\partial z}=-\frac{\partial w}{\partial x}+\mu\left(\frac{\partial^2 w}{\partial x^2}+\frac{\partial^2 w}{\partial y^2}+\frac{\partial^2 w}{\partial z^2}\right) \tag{5.8}$$

The 3D steady-state energy conservation equation applicable only for the IC chip domain is given in Equation 5.9.

$$u\frac{\partial T}{\partial x}+v\frac{\partial T}{\partial y}+w\frac{\partial T}{\partial z}=\alpha\left(\frac{\partial^2 T}{\partial x^2}+\frac{\partial^2 T}{\partial y^2}+\frac{\partial^2 T}{\partial z^2}\right)+Q_g \tag{5.9}$$

5.6.2 Boundary Conditions

The boundary conditions applied to the computational model are expressed next. Inlet boundary condition:

At inlet (X = 0), T = T∞ = 303.15 K, u = 4.5 m/s and 8 m/s, v = w = 0
Outlet boundary condition: At outlet (X = L), P = P_∞

The lateral boundary condition of the test section is considered to be adiabatic,

$$\frac{\partial T}{\partial y}=\frac{\partial T}{\partial z}=0$$

5.6.3 Mesh Independence Study

The selection of proper mesh plays a vital role in determining the final solution to the problem. Normally, for the numerical investigation, around 80% of the time is spent in determining the optimum mesh size for which the final solution doesn't vary with the number of nodes and elements. In the present study, the mesh independence study is carried out to minimize the error and to have a consistent solution. The optimum mesh element is

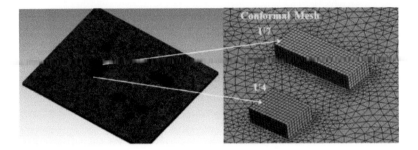

FIGURE 5.16   Optimal mesh profile for the simulation of optimal configuration ($\lambda = 1.68806$).

chosen for the present study having 1,468,088 elements and 256,424 nodes with less maximum skewness of 0.8413. Increasing the mesh elements above these values has a negligible effect on the final solution of the problem and the time taken to solve the problem is also minimum. The maximum temperature of the IC chips does not vary with the mesh elements above the optimum grid size. The optimal mesh profile selected for the simulation of the optimal configuration ($\lambda = 1.6880$) is shown in Figure 5.16 and the results of the mesh independence study are shown in Figure 5.17. The convergence criterion used for the mass and momentum equation is 10^{-3} and for the energy equation, it is 10^{-6}, respectively.

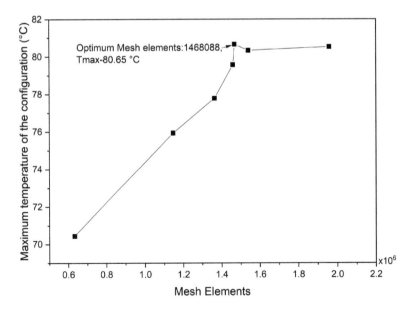

FIGURE 5.17   Results of the mesh independence study.

5.7 NUMERICAL ANALYSIS FOR THE IC CHIP'S TEMPERATURE UNDER THE DIFFERENT SUBSTRATE BOARD ORIENTATIONS

The numerical simulations are carried out under the steady-state laminar forced convection for the different configurations as mentioned in Table 5.1. The actual boundary conditions as that of the experimental analysis are used for the numerical model. The simulations are then validated with the experiments, as shown in Figure 5.18. It is seen that both the results agree well within an error band of 1.90–7.66%. The small deviation in the numerical results is due to the idealistic condition in the computational model and neglecting the assumptions that are made for the realistic model.

Figure 5.19 shows the comparison of temperature excess ($T_{IC} - T_\infty$) for all the IC chips of different configurations for both the numerical and the experimental analysis. It is seen that both the results are in strong agreement with a deviation of ±12.35%.

The temperature contour and the velocity profiles for the optimal configuration ($\lambda = 1.68806$) are shown in Figures 5.20 and 5.21, respectively.

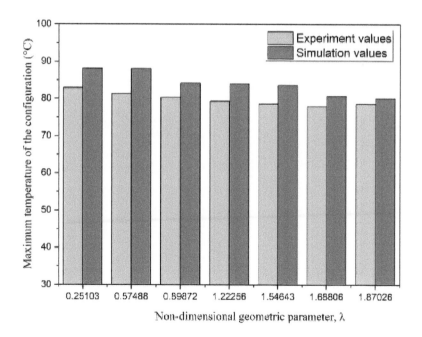

FIGURE 5.18 Validation of simulation and experiment results for configuration maximum temperature.

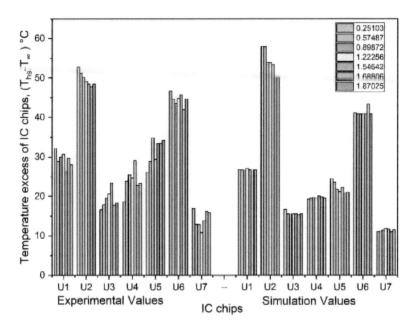

FIGURE 5.19   Validation of simulation and experiment results for IC chip temperatures of different configurations.

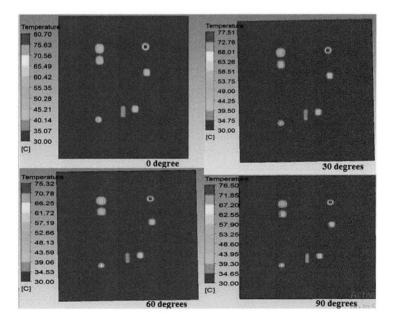

FIGURE 5.20   Temperature contours for the IC chips of the optimal configuration ($\lambda = 1.68806$) under different substrate board orientations.

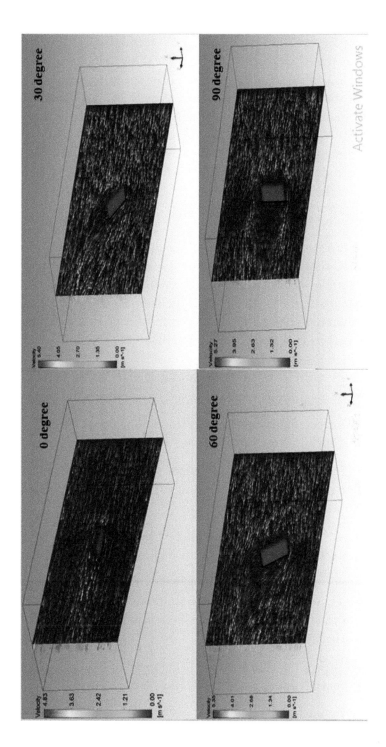

FIGURE 5.21   Velocity profile for the IC chips of the optimal configuration ($\lambda = 1.68806$) under different substrate board orientations.

It is observed that the space available to carry away the heat from the IC chips is more for the 0° (horizontal) and 30° orientation of the substrate board. Hence, the IC chips are cooled fast for the 0°and 30° orientations of the board as compared to the other one. The space to carry the heated air is less for the 60° and 90° orientations of the substrate board because there is a recirculation of air on the top surface of the board and this leads to the formation of eddies. Due to this, there is an increase in the thermal interaction among the IC chips leading to an increase in their temperature. Therefore, the cooling effect of the IC chips is less for the 60° and 90° orientations of the substrate board.

5.8 CONCLUSIONS

Steady-state experimental and numerical analysis is carried out on seven asymmetric IC chips mounted at different positions on a substrate board under the laminar forced convection. The substrate board is oriented at four different angles of 0°, 30°, 60°, and 90°. The objective is to determine the optimal configuration for the arrangement of these IC chips. The following conclusions are brought down from the study.

1. The temperature of the IC chips is least for the 30° substrate board and the vertical board attains the maximum temperature.

2. The optimal configuration is determined using the λ; the maximum temperature of the configuration decreases with the increase in λ.

3. There is a temperature drop of about 5°C between the lower and upper extreme configurations leading to better thermal management and ultimately cools the chips faster.

4. Air velocity is crucial for the cooling of the IC chips; there is a temperature drop of 1.42–17.33% for the higher velocity (8 m/s) as compared to the lower one (4.5 m/s).

5. Different correlations are proposed to predict the temperature of the IC chips in terms of their size, position on the board, and substrate board orientation.

6. The hybrid optimization strategy (combined ANN-GA) can predict the optimal configuration of the IC chips more accurately.

7. The optimum configuration ($\lambda = 1.68806$) obtained from the ANN-GA strategy confirms that the IC chips with higher heat flux

and temperature should be placed either at the upper edge and lower edge of the substrate board leading to negligible thermal interaction among the other chips.

8. The numerical simulation results are in strong agreement with the experiments with an error band of $\pm 12.35\%$.

9. The proposed results provide a good insight into the thermal design engineers to identify the positioning of IC chips on the substrate board to increase their reliability and working cycle.

CHAPTER 6

Numerical and Experimental Investigations of Paraffin Wax-Based Mini-Channels for the Cooling of IC Chips

6.1 INTRODUCTION

In the previous chapters, the studies are conducted to determine the optimal distribution of the integrated circuit (IC) chips under the mixed convection and laminar forced convection heat transfer mode driven by the numerical and experimental data, respectively. The temperature of the IC chips is further minimized using the phase change material (PCM) based mini-channels extruded at the periphery of the substrate board in which the conjugate heat transfer from the IC chips has played a vital role. The present chapter deals with the transient experimental and numerical investigations carried out for the cooling of the seven asymmetric IC chips under the natural convection heat transfer mode using Paraffin wax-based mini-channels. Paraffin wax is filled inside the protrude mini-channels (fabricated on the periphery of the substrate board) and the IC chips are placed adjacent to the channels. As discussed under Section 1.4.2, Paraffin wax is the preferred PCM for the thermal management of electronic components, as it has a high latent heat of fusion, low value of thermal conductivity, reliable charging and discharging cycle, a wide range of melting

DOI: 10.1201/9781003188506-6

point, and it is also non-reactive, and non-corrosive. The numerical analysis is carried out using the different PCMs inside the mini-channels using their variable volume of fluid (VoF). The objective is to increase the performance and working cycle of the IC chips using Paraffin wax based mini-channels. For a better understanding, the chapter is classified into the following sections:

- Experimental set-up and methodology

- Effect of Paraffin wax-based mini-channels on the temperature of the IC chips

- Numerical framework

6.2 EXPERIMENTAL SET-UP

The transient experiments are conducted for the seven asymmetric IC chips placed adjacent to Paraffin wax-based mini-channels fabricated at the periphery of the substrate board. The IC chips are supplied with variable heat fluxes (leading to non-uniform volumetric heat generation ranging from 4×10^6–10×10^6 W/m^3). The design of the substrate board, IC chips, and the fabrication of the mini-channels are already detailed under Section 3.3. The schematic diagram of the experimental facility using Paraffin wax (T_m: 48.28°C–54.21°C) based on mini-channels is shown in Figure 6.1. The procedure for conducting the transient experiments is detailed under Section 3.5.2.

6.3 RESULTS AND DISCUSSION

The transient experiments are conducted on the seven asymmetric IC chips placed adjacent to the mini-channels under the natural convection heat transfer mode. The experiments are conducted by supplying four different volumetric heat-generation values to the IC chips under two set point parameters (SPP) – the first one (SPP-1) is the time taken by the IC chips to reach the set point temperature (SPT) of 85°C without using the PCM-based mini-channels (WPMC), and the second one (SPP-2) is the time taken for the PCM to reach its 90% melting inside the mini-channels (MPMC). The 90% PCM melting (not full melting) is considered so that the liquid PCM does not flow outside the channel as the channels are open at both ends. The objective of the current study is to increase the working cycle of the IC chips using the PCM-based mini-channels (PMC) under the case of variable heat flux distribution.

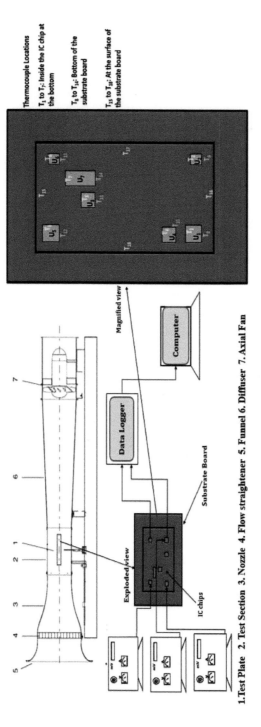

FIGURE 6.1   Schematic diagram of the experimental set-up.

1. Test Plate 2. Test Section 3. Nozzle 4. Flow straightener 5. Funnel 6. Diffuser 7. Axial Fan

6.3.1 Temperature Variation of IC Chips without PCM-Based Mini-Channels

Figure 6.2 shows the temperature variation of the IC chips of a particular configuration without the PCM-based mini-channels (WPMC) under four different cases (case 1: 10×10^6 W/m^3, case 2: 8×10^6 W/m^3, case-3: 6×10^6 W/m^3, and case 4: 4×10^6 W/m^3). The maximum temperature of the configuration (arrangement of the seven IC chips) reaches the set point temperature (SPT) of 85°C in a short period, i.e. between 700 s and 2,065 s under the natural convection heat transfer mode for the earlier cases. The SPT has reached in 700 s for case 1 (lowest volumetric heat generation) for the IC chip U_6 (smallest chip), while for the remaining cases, it is reached in 944 s, 1,292 s, and 2,065 s, respectively, for the IC chip U_3.

Most portable electronic devices rely on natural convection heat transfer mode and due to miniaturization, the space available for their cooling is also very less. The objective of the current study is to increase the working cycle of IC chips under the natural convection heat transfer mode using the PMC. Hence, the aluminium mini-channels are fabricated and flushed inside the periphery of the substrate board in which Paraffin wax is filled. The arrangement for the PMC is shown in Figure 6.1.

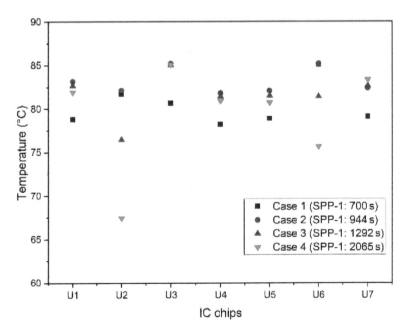

FIGURE 6.2 Temperature variation of IC chips without PCM-based mini-channels (WPMC).

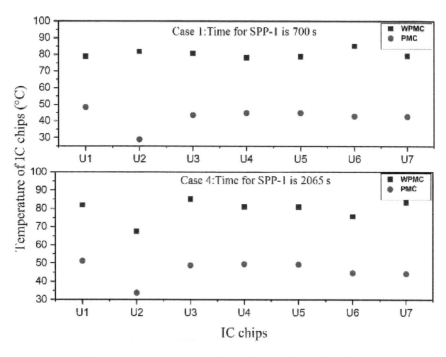

FIGURE 6.3   Temperature variation of IC chips with and without PCM-based mini-channels to reach the SPP-1.

Figure 6.3 shows the temperature variation of the IC chips to reach the SPT for two different cases (1 and 4) with and without the PMC. The use of Paraffin wax inside the mini-channels has resulted in a significant temperature drop (37.34–45.79%) from the IC chips.

6.3.2 Temperature Variation of IC Chips for Case 1 with and without the PCM-Based Mini-Channels

Figure 6.4 shows the temperature variation of the IC chips with and without the PMCs for case 1 under the condition of SPP-2 (90% melting of PCM). For case 1, the seven asymmetric IC chips are supplied with a constant volumetric heat generation of 10×10^6 W/m^3.

It is observed that the PCM-based left mini-channel (LMC) has reached 90% melting of PCM at t = 945 s and by this time, the right mini-channel (RMC) has reached only 31.98% melting of PCM. The temperatures of the IC chips U_1, U_3, U_4, and U_5 placed adjacent to the LMCs are 55.5°C, 50.3°C, 51.9°C, and 51.8°C, respectively. The temperatures of these IC chips are compared without the WPMC, and it is seen that their working cycle has

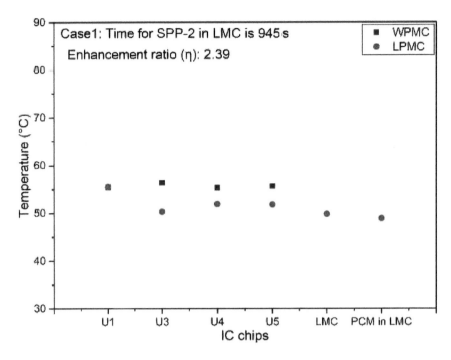

FIGURE 6.4 Temperature variation of IC chips for case 1 with and without the PCM-based left mini-channels (LMCs) under the condition of SPP-2.

been enhanced by 2.39 times. The power supply to these four IC chips is then cut OFF, as these chips have already reached the condition of SPP-2 (90% melting of PCM).

However, the power supply to the IC chips U_2, U_6, and U_7 placed adjacent to the RMC are continued further, and the condition of SPP-2 for these chips is reached in t = 2,113 s, as shown in Figure 6.5. The working cycle of these IC chips is enhanced by 3.01 times as compared to WPMC. From Figures 6.4 and 6.5, it is clear that the working cycle of the IC chips has been enhanced by 2.39–4.31 times to reach the condition of SPP-2 for case 1.

Figure 6.6 shows the charging and discharging phases of the IC chips placed adjacent to the LMCs and RMCs for case 1 under the condition of SPP-2. It shows the total operating time of the IC chips U_1, U_3, U_4, and U_5 (placed adjacent to the LMC) as 2,113 s (charging time: 945 s, discharging time: 1,168 s). The power supply to these IC chips is then cut OFF, as the PCM in the left channel has reached 90% melting. However, during this time, the IC chips U_2, U_6, and U_7 (placed adjacent to the right channel) are still under the charging phase. The temperature of these IC chips follows the same trend

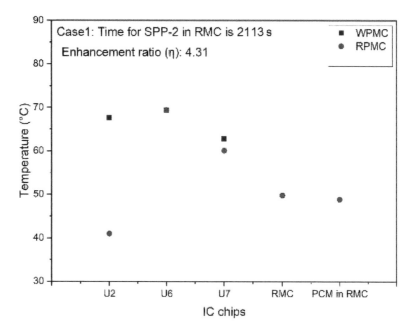

FIGURE 6.5   Temperature variation of IC chips for case 1 with and without the PCM-based right mini-channels (RMCs) under the condition of SPP-2.

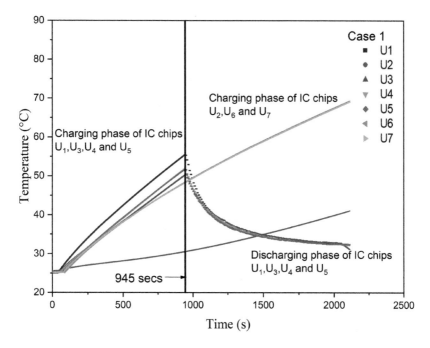

FIGURE 6.6   Temperature variation of IC chips for case 1 with charging and discharging phase under the condition of SPP-2.

of increasing temperature with respect to time. Therefore, the PCM-based right channel reaches the condition of 90% melting (SPP-2) in 2,113 s.

6.3.3 Temperature Variation of IC Chips for Case 4 with and without the PCM-Based Mini-Channels

The temperature variation of the IC chips for case 4 with and without the PMCs is shown in Figure 6.7.

The temperature of the IC chips is reduced using the PMC and compared to the WPMC. The PCM in the LMC has reached 90% melting in $t = 2,139$ s and during which the temperatures of the IC chips U_1, U_3, U_4, and U_5 are 52°C, 49.7°C, 50.4°C, and 50.1°C, respectively. The temperature of these IC chips is then compared without the WPMC, and it is seen that their working cycle is enhanced by 2.32 times.

Figure 6.8 shows the temperature variation of the IC chips U_2, U_6, and U_7 placed adjacent to the RMC. In this case, the PCM attains 90% melting at $t = 4,534$ s and thus enhances the working cycle of the IC chips by 3.90 times. This indicates that the temperature of the IC chips has reduced, thus increasing their working cycle using the PMC. The same trend is observed for cases 2 and 3 as well.

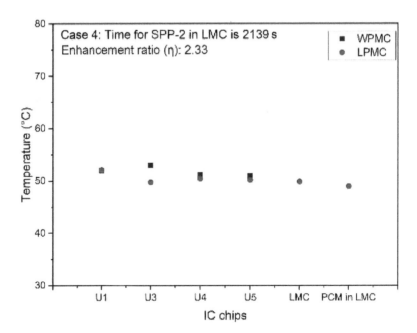

FIGURE 6.7 Temperature variation of IC chips for case 4 with and without the PCM-based LMCs under the condition of SPP-2.

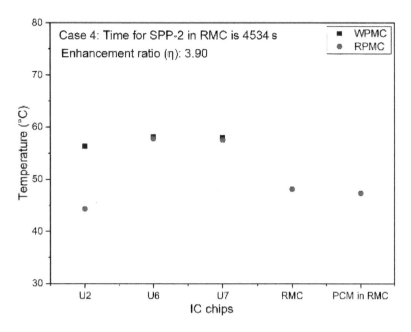

FIGURE 6.8 Temperature comparison of IC chips with and without PCM-based RMCs under SPP-2 for case 4.

Table 6.1 shows the temperature of the IC chips at which the PCM in the LMC attains 90% melting in comparison to WPMC. The working cycle of the IC chips U_1, U_3, U_4, and U_5 placed adjacent to the LMC has been enhanced by 2.33–2.57. The power supply to these IC chips is then cut OFF after reaching the condition of SPP-2 (90% PCM melting) inside the left

TABLE 6.1 Temperature of IC Chips to Reach the Condition of SPP-2 for the Left Mini-Channel (LMC) Case

Different Cases	Time, s	Working Cycle Enhancement	U_1	U_2	U_3	U_4	U_5	U_6	U_7	RMC	PCM-R	LMC	PCM-L
Case 1	945	2.39	55.5	30.5	50.3	51.9	51.8	48.4	48.2	33.1	32.8	49.8	48.9
Case 1 (WPMC)	396	–	55.4	60.7	56.3	55.3	55.6	61.8	55.6	–	–	–	–
Case 2	1242	2.57	54.9	29.9	50.5	51.8	51.9	47.6	47.2	34	33.5	49.9	48.7
Case 2 (WPMC)	483	–	55	59.6	56.2	54.2	54.3	59.1	53.6	–	–	–	–
Case 3	1679	2.73	53.5	32.3	50.1	51.1	51	45.7	45.3	34.3	33.8	49.7	48.7
Case 3 (WPMC)	615	–	53.1	55.9	54.3	52.3	52.3	55.8	51.8	–	–	–	–
Case 4	2139	2.33	52	34.1	49.7	50.4	50.1	45.3	45	37	36.5	49.8	48.9
Case 4 (WPMC)	920	–	52	51.8	52.9	51.1	50.9	52.2	50.6	–	–	–	–

channel. These IC chips are then cooled down and lose heat under the natural convection heat transfer mode because the power supply to the IC chips U_2, U_6, and U_7 is still under the charging phase till the PCM is in the right channels attains the condition of SPP-2.

6.3.4 Temperature Variation of IC Chips for Cases with PCM-Based Mini-Channels

As mentioned under Section 6.2, the experiments are conducted to reach the condition of 90% PCM melting for both the LMCs and RMCs. For all the cases, the PCM-based LMC reaches the condition much earlier as compared to the RMC. Therefore, the IC chips U_1, U_3, U_4, and U_5 are under the discharging process after the condition has reached the LMC. Figure 6.9 shows the charging and discharging of the PCM-based LMC with time.

Figure 6.10 shows the volume fraction variation of the PCM during the charging phase for the RMC case. It depicts that the right channel reaches the condition of 90% PCM melting at the latter stage reducing the temperature of the IC chips U_2, U_6, and U_7.

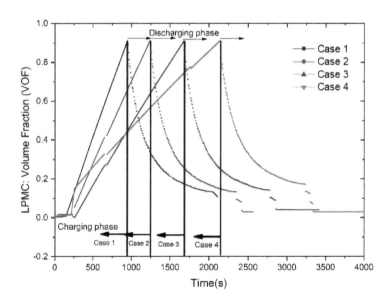

FIGURE 6.9 Temperature variation of IC chips for all cases with PCM-based mini-channels (PMCs).

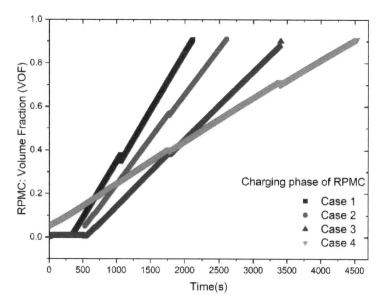

FIGURE 6.10 Variation of PCM volume fraction in the RMC with time.

Table 6.2 gives the temperature of the IC chips U_2, U_6, and U_7 under the charging phase and the IC chips U_1, U_3, U_4, and U_5 under the discharging phase till the condition of 90% PCM melting has reached in the RMC.

TABLE 6.2   Temperature of IC Chips to Reach the Condition SPP-2 for the Right Mini-Channel (RMC) Case

Different Cases	Time, s	Working Cycle Enhancement	U_1	U_2	U_3	U_4	U_5	U_6	U_7	RMC	PCM-R	LMC	PCM-L
Case 1	2,113	4.31	32	41	31	32	32	69	69	49.8	48.9	28.7	27.6
Case 1 (WPMC)	490	–	63	68	64	62	63	69	63	–	–	–	–
Case 2	2,607	4.45	32	37	27	27	32	66	65	49.9	48.7	25.8	25.7
Case 2 (WPMC)	586	–	62	65	63	61	61	66	60	–	–	–	–
Case 3	3,414	4.63	26	44	26	27	26	61	61	49.6	48.8	26.1	26
Case 3 (WPMC)	738	–	59	61	60	58	58	61	58	–	–	–	–
Case 4	4,534	3.9	26	44	27	27	27	58	58	49.9	48.7	27	25.8
Case 4 (WPMC)	1,162	–	59	56	60	58	58	58	58	–	–	–	–

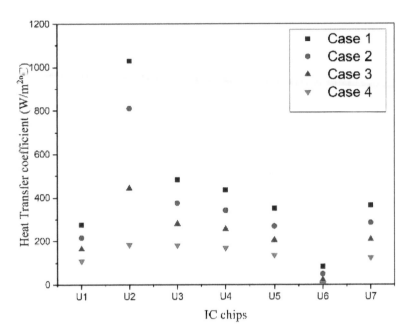

FIGURE 6.11   Heat transfer coefficient variation for all IC chips for different cases.

6.3.5 Convective Heat Transfer Coefficient Variation for Cases with PCM-Based Mini-Channels (PMCs)

The convective heat transfer coefficient plays an important role in enhancing the heat transfer rate of the IC chips which depicts their degree of cooling. Figure 6.11 shows the variation of the convective heat transfer coefficient of all the IC chips for four different cases. The heat transfer coefficient for case 1 is higher as compared to case 4. The IC chip U_2 has a higher heat transfer co-efficient for all cases, as it has the smallest size among the other IC chips. It suggests that the smallest size IC chip U_2 attains the highest temperature and eventually cooled down using the PCM, thus increasing its working cycle.

6.3.6 Correlation

A correlation is developed for the non-dimensional maximum temperature excess, θ of the IC chips in terms of the Fourier number, Fo under all four cases. Fourier number is the non-dimensional parameter used to predict the temperature of the IC chips with the response to the PCM melting. The relation is given in Equation 6.1.

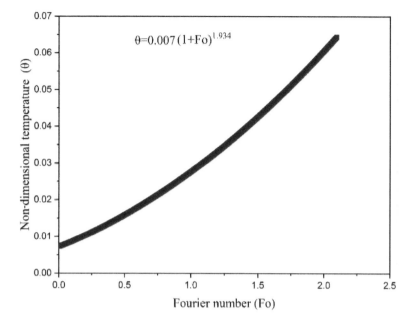

FIGURE 6.12   Variation of the non-dimensional maximum temperature excess of IC chips with Fourier number.

$$\theta = 0.007\left(1+\text{Fo}\right)^{1.934} \qquad (6.1)$$

Equation 6.1 has a regression coefficient of 0.92, the RMS error is 0.0013422, and is valid for $0.009802 \leq \text{Fo} \leq 2.096784$, $0.000579 \leq \theta \leq 0.058769$. Figure 6.12 shows the variation of the non-dimensional maximum temperature excess of the IC chips with their Fourier number. The higher value of the Fourier number signifies the degree of heat conducted in the PCM relative to the PCM energy storage.

6.4 NUMERICAL SIMULATION OF PCM-BASED MINI-CHANNELS UNDER NATURAL CONVECTION

Transient numerical simulations are carried out for the PMC for all the previously mentioned four cases under natural convection. The methodology adopted for the numerical simulation is already detailed under Section 6.2. Figure 6.13 shows variation of the maximum temperature of IC chips with time for case 1.

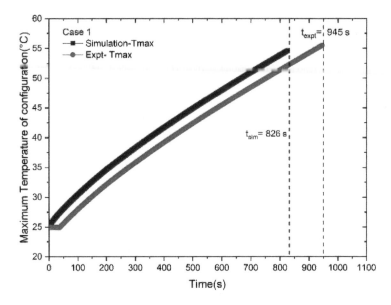

FIGURE 6.13 Validation of the numerical model with the experimental analysis.

The trend of rise in the maximum temperature of the configuration for case 1 under both the numerical and experimental analysis is similar to a temperature difference of 4–5°C. However, the time is taken to reach the 90% PCM melting in the LMC for both the numerical and experimental analysis is 826 s and 945 s, respectively. It indicates that the experimental and numerical results are in strong agreement with each other. Similar trends are observed for all the other cases.

Figures 6.14 and 6.15 show the temperature contours and the melt front for case 1 at the time steps of 100 s. It suggests the temperature distribution among the IC chips and the contours suggest their melting rate with time. Figures 6.16 and 6.17 show the validation of the numerical model with the experimental analysis for the PCM-based LMC under the condition of 90% melting. It is seen that the temperature of all the IC chips is in strong agreement with the experiments. In the present experimental and numerical study, Paraffin wax is used as the PCM inside the mini-channels. The transient analysis is carried out under the natural convection heat transfer mode.

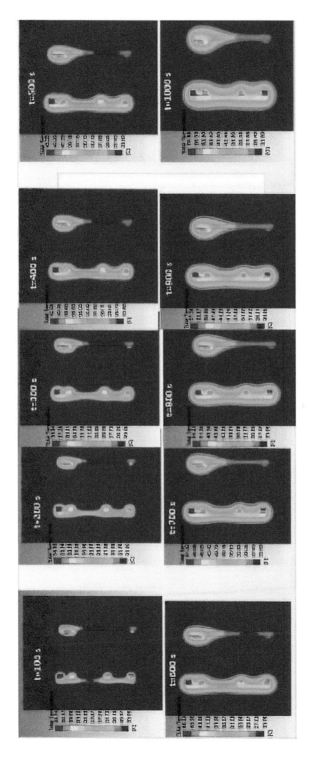

FIGURE 6.14   Temperature contours of the IC chips with time for case 1.

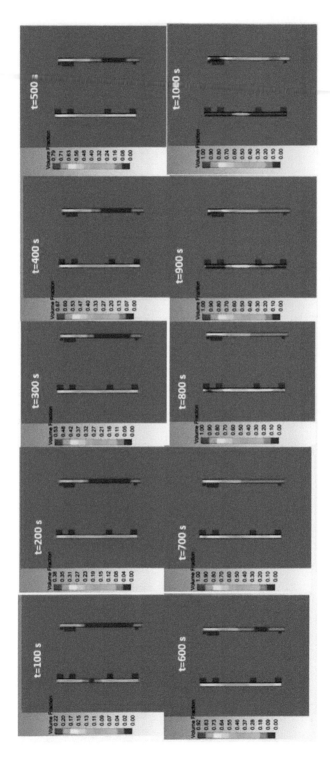

FIGURE 6.15 Variation of melt fraction of the PCM with time for case 1.

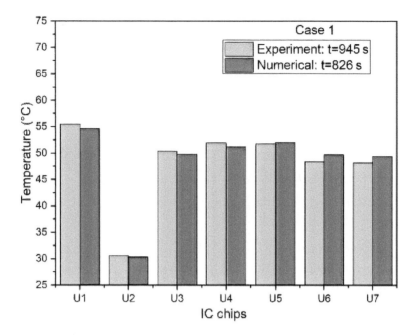

FIGURE 6.16   Validation of the numerical model with the experiments for case 1.

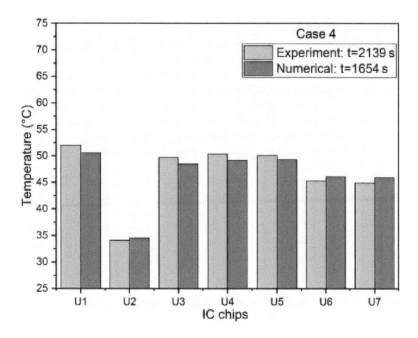

FIGURE 6.17   Validation of the numerical model with the experiments for case 4.

6.5 CONCLUSIONS

The transient experimental and numerical simulations are carried out under the natural convection heat transfer mode for the cooling of the seven asymmetric IC chips mounted on a substrate board using Paraffin wax-based mini-channels. The following conclusions are laid down from the study:

1. The IC chips have reached the SPT in 700 s to 2,065 s for the different cases 1–4, respectively of the WPMC.

2. The working cycle of the IC chips has been enhanced using the PCM-based mini-channels (PMC).

3. The working cycle of IC chips U_1, U_3, U_4, and U_5 has been enhanced by 2.33–2.73 times, and the IC chips U_2, U_6, and U_7 have been enhanced by 3.9–4.63 times in comparison to WPMC.

4. Numerical results are validated with the experiments and the temperature difference is found to be only 4–5°C, which suggests that both the results are in strong agreement with each other.

5. Paraffin wax has led to the IC chip temperature drop of 4–12%.

6. The temperature of the IC chips has dropped significantly using the n-eicosane (lower melting point). However, the maximum heat storage has occurred using Paraffin wax.

7. The cooling (discharging) cycle of the IC chips is much faster as compared to their heating (charging) one.

8. Paraffin wax has reduced the temperature of the IC chips significantly and ultimately cools these faster.

CHAPTER 7

Conclusions and Scope for Future Work

7.1 INTRODUCTION

The present study deals with the experimental and numerical analysis on the seven asymmetric high heat-generating integrated circuit (IC) chips mounted at different positions on a switch-mode power supply (SMPS) board cooled using air and phase change material (PCM) under the different heat transfer modes. The application domain-specific module SMPS board is considered for the analysis. The objective is to decide the optimal configuration for the arrangement of these seven IC chips on the board. The maximum temperature excess must be the minimum among all the other possible configurations for the optimal configuration. This is determined by introducing a non-dimensional heuristic geometric distance parameter (λ). The actual IC chips and the SMPS board are mimicked with aluminium and Bakelite, respectively. The domain considered for the analysis of the SMPS board mounted with the IC chips is taken into consideration with the practical application. The brief highlights for analyzing the seven non-identical IC chips mounted on the SMPS board are reported next in chronological order.

1. Initially, the SMPS board and the seven asymmetric IC chips are selected. The numerical analysis is carried out using the ANSYS Icepak (V16.0) for the IC chips cooled using air under mixed convection. Based on the numerical analysis, an empirical correlation

DOI: 10.1201/9781003188506-7

is developed for the non-dimensional maximum temperature excess (θ) of the IC chips in terms of their non-dimensional position parameter (λ). A numerical data-driven combined artificial neural network-genetic algorithm (ANN-GA) based technique is then employed to determine the global optimal position of the IC chips.

2. Steady-state experiments are then conducted to study the substrate board orientation effect on the temperature of the IC chips cooled using air under the laminar forced convection heat transfer. Based on the experimental results, an empirical correlation is developed for the non-dimensional maximum temperature excess (θ) of the IC chips in terms of their λ, size (S), and non-dimensional substrate board orientation (φ). The experimental data-driven combined ANN-GA-based technique is then developed to determine the global optimal position of IC chips.

3. The optimal configuration obtained using the numerical data-driven combined ANN-GA-based technique is further analyzed numerically to increase the working cycle of the IC chips using the PCM-based mini-channels. The role of different PCMs in cooling the IC chips of the optimum configuration is also studied.

4. The transient experimental and numerical analysis are then conducted under the natural convection heat transfer mode using Paraffin wax-based mini-channels fabricated on the periphery of the substrate board with the IC chips placed adjacent to it. The numerical results are validated with the experiments.

The book started with an introduction in identifying the need for thermal management in the electronic devices, different techniques available for their cooling with an emphasis on the PCMs for the IC chips cooling. The increasing demand for the use of PCM in electronic cooling is also discussed in Chapter 1. This is followed by Chapter 2, which presents a detailed literature review pertinent to the cooling of IC chips using both air and PCM. From the conclusion of the literature, the objectives of the book are laid down. Chapter 3 discusses the design and selection of the IC chips and the SMPS board. This chapter also highlights the details about the experimental facility used in the book along with the methodology to carry out the experiments under the different heat transfer modes using both air and PCM. Chapter 4 proposes a methodology for determining the optimal configuration of the seven asymmetric IC chips mounted at different positions on an SMPS board cooled using air under the mixed convection heat transfer mode. The numerical data-driven combined ANN-GA-based technique is employed to determine the optimal configuration. The steady-state experimental and numerical analysis to study the effect of substrate board orientation on the temperature of the IC chips under the laminar forced convection heat transfer mode is discussed in Chapter 5. This chapter also determines the optimal configuration of the IC chips using the experimental data-driven combined ANN-GA-based technique. Chapter 6 highlights the experimental and numerical analysis for Paraffin wax-based mini-channels under the laminar natural convection heat transfer mode. The role of different PCMs for the cooling of the IC chips is also analyzed.

7.2 MAJOR CONCLUSIONS OF THE PRESENT STUDY

Based on the detailed numerical, optimization, and experimental analysis for the problem under consideration, the major conclusions of the study are highlighted next:

1. The temperature of the IC chips is a strong function of their size, position on the substrate board, and input heat flux value.

2. The size of the IC chips and their positioning on the board are characterized by the non-dimensional position parameter, λ. The highest λ signifies a minimum of the maximum configuration (arrangement of seven IC chips) temperature and thus leads to the optimal one.

3. The smallest size IC chip, U_2 with a high heat dissipation rate leads to the maximum temperature in the configuration. However, the temperature variation for the low-powered IC chips U_3, U_4, and U_7 are very small (close to ambient).

4. The numerical data-driven combined ANN-GA-based strategy predicts the optimal configuration of the IC chips more accurately as compared to the conventional one.

5. The optimal configuration obtained from the ANN-GA-based strategy confirms that the IC chips dissipating more heat must be placed at the substrate bottom.

6. The substrate board orientation has a significant effect on the cooling of the IC chips which is a minimum for 30° substrate boards and follows the relation 30° < 0° < 60° < 90°.

7. Different correlations are proposed to predict the temperature of the IC chips in terms of their size (S), position on the board (λ), and non-dimensional substrate board orientation (φ).

8. The optimum configuration ($\lambda = 1.68806$) obtained from the experimental data-driven combined ANN-GA-based strategy confirms that the IC chips with higher heat flux and temperature should be placed either at the upper edge or lower edge of the substrate board leading to less thermal interaction among the other chips.

9. The PCM plays a significant role to reduce the temperature of the IC chips (up to 13.66%) and ultimately cools those faster.

10. The working cycle of the IC chips U_1, U_3, U_4, and U_5 is increased by 2.33 times to 2.73 times, and the IC chips U_2, U_6, and U_7 are increased by 3.9 times to 4.6 times in comparison to without Paraffin wax-based mini-channels (WPMC).

11. Paraffin wax has led to the IC chip temperature drop of 4–12%.

12. The temperature of the IC chips is dropped significantly using the n-eicosane (lower melting point). However, maximum heat is stored using Paraffin wax.

13. The proposed results provide a good insight into the thermal design engineers to identify the positioning of IC chips on the substrate board to increase their reliability and working cycle.

7.3 SCOPE FOR FUTURE WORK

The study presented in the book can be extended to more number of IC chips placed along with the other components on the substrate board. The cooling techniques used in the present study along with the optimization strategy can be extended in carrying out the experimental and numerical analysis as highlighted next:

1. PCM-based heat sink can be placed between the different components of the SMPS board along with the IC chips.

2. Thermal interface material (TIM) can be attached to the IC chips to increase their working cycle.

3. The current work can be extended to the battery thermal management systems using the PCM.

4. The micro-channel-based liquid cooling system can be employed at the bottom of the thermally conductive material of the SMPS board.

5. The cold plate can be used for the cooling of the high heat-generating IC chips to increase their working cycle.

References

Abokersh, M.H., Osman, M., El-Baz, O., El-Morsi, M. and Sharaf, O. (2018) 'Review of the phase change material (PCM) usage for solar domestic water heating systems (SDWHS)', *International Journal of Energy Research*. Wiley Online Library, **42**(2), pp. 329–357.

Ahmed, S.E., Mansour, M.A., Hussein, A.K. and Sivasankaran, S. (2016) 'Mixed convection from a discrete heat source in enclosures with two adjacent moving walls and filled with micropolar nanofluids', *Engineering Science and Technology, an International Journal*. Elsevier, **19**(1), pp. 364–376.

Ajmera, S.K. and Mathur, A.N. (2015) 'Experimental investigation of mixed convection in multiple ventilated enclosure with discrete heat sources', *Experimental Thermal and Fluid Science*. Elsevier Inc., **68**, pp. 402–411.

Ali, H.M. (2018) 'Experimental investigation on paraffin wax integrated with copper foam-based heat sinks for electronic components thermal cooling', *International Communications in Heat and Mass Transfer*. Elsevier, **98**, pp. 155–162.

Ali, H.M., Ashraf, M.J., Giovannelli, A., Irfan, M., Irshad, T.B., Hamid, H.M., Hassan, F. and Arshad, A. (2018) 'Thermal management of electronics: An experimental analysis of triangular, rectangular and circular pin-fin heat sinks for various PCMs', *International Journal of Heat and Mass Transfer*. Elsevier, **123**, pp. 272–284.

Aminossadati, S.M. and Ghasemi, B. (2009) 'A numerical study of mixed convection in a horizontal channel with a discrete heat source in an open cavity', *European Journal of Mechanics-B/Fluids*. Elsevier, **28**(4), pp. 590–598.

Amirouche, Y. and Bessaih, R. (2012) 'Three-dimensional numerical simulation of air cooling of electronic components in a vertical channel', *Fluid Dynamics & Materials Processing*. Tech Science Press, **8**(3), pp. 295–309.

Arshad, A., Ali, H.M., Khushnood, S. and Jabbal, M. (2018) 'Experimental investigation of PCM based round pin-fin heat sinks for thermal management of electronics: Effect of pin-fin diameter', *International Journal of Heat and Mass Transfer*. Elsevier, **117**, pp. 861–872.

Ashraf, M.J., Ali, H.M., Usman, H. and Arshad, A. (2017) 'Experimental passive electronics cooling: Parametric investigation of pin-fin geometries and efficient phase change materials', *International Journal of Heat and Mass Transfer*. Elsevier, **115**, pp. 251–263.

Aydin, O. and Yang, W. (2000) 'Natural convection in enclosures with localized heating from below and symmetrical cooling from sides', *International Journal of Numerical Methods for Heat & Fluid Flow*. Emerald, **10**(5), pp. 518–529.

Baby, R. and Balaji, C. (2012) 'Experimental investigations on phase change material based finned heat sinks for electronic equipment cooling', *International Journal of Heat and Mass Transfer*. Elsevier, **55**(5–6), pp. 1642–1649.

Baby, R. and Balaji, C. (2013) 'Thermal optimization of PCM based pin fin heat sinks: An experimental study', *Applied Thermal Engineering*. Elsevier, **54**(1), pp. 65–77.

Bakkas, M., Amahmid, A. and Hasnaoui, M. (2008) 'Numerical study of natural convection heat transfer in a horizontal channel provided with rectangular blocks releasing uniform heat flux and mounted on its lower wall', *Energy Conversion and Management*. Elsevier, **49**(10), pp. 2757–2766.

Bazylak, A., Djilali, N. and Sinton, D. (2006) 'Natural convection in an enclosure with distributed heat sources', *Numerical Heat Transfer, Part A: Applications*. Taylor & Francis, **49**(7), pp. 655–667.

Bessaih, R. and Kadja, M. (2000) 'Turbulent natural convection cooling of electronic components mounted on a vertical channel', *Applied Thermal Engineering*. Elsevier, **20**(2), pp. 141–154.

Bhowmik, H. and Tou, K.W. (2005) 'Experimental study of transient natural convection heat transfer from simulated electronic chips', *Experimental Thermal and Fluid Science*. Elsevier, **29**(4), pp. 485–492.

Bhowmik, H., Tso, C.P., Tou, K.W. and Tan, F.L. (2005) 'Convection heat transfer from discrete heat sources in a liquid-cooled rectangular channel', *Applied Thermal Engineering*. Elsevier, **25**(16), pp. 2532–2542.

Bondareva, N.S., Buonomo, B., Manca, O. and Sheremet, M.A. (2019) 'Heat transfer performance of the finned nano-enhanced phase change material system under the inclination influence', *International Journal of Heat and Mass Transfer*. Elsevier, **135**, pp. 1063–1072.

Chaurasia, N.K., Gedupudi, S. and Venkateshan, S.P. (2019) 'Studies on three-dimensional mixed convection with surface radiation in a rectangular channel with discrete heat sources', *Heat Transfer Engineering*. Taylor & Francis, **40**, 66–80.

Chen, S. and Liu, Y. (2002) 'An optimum spacing problem for three-by-three heated elements mounted on a substrate', *Heat and Mass Transfer*. Springer, **39**(1), pp. 3–9.

Choi, C.Y. and Ortega, A. (1993) 'Mixed convection in an inclined channel with a discrete heat source', *International Journal of Heat and Mass Transfer*. Elsevier, **36**(12), pp. 3119–3134.

Chuang, S., Chiang, J.S. and Kuo, Y.M. (2003) 'Numerical simulation of heat transfer in a three-dimensional enclosure with three chips in various position arrangements', *Heat Transfer Engineering*. Taylor & Francis, **24**(2), pp. 42–59.

Cui, Y., Liu, C., Hu, S. and Yu, X. (2011) 'The experimental exploration of carbon nanofiber and carbon nanotube additives on the thermal behavior of phase change materials', *Solar Energy Materials and Solar Cells*. Elsevier, **95**(4), pp. 1208–1212.

Da Silva, A.K., Lorente, S. and Bejan, A. (2004) 'Optimal distribution of discrete heat sources on a plate with laminar forced convection', *International Journal of Heat and Mass Transfer*. Elsevier, **47**(10–11), pp. 2139–2148.

Demirbas, M.F. (2006) 'Thermal energy storage and phase change materials: An overview', *Energy Sources, Part B: Economics, Planning, and Policy*. Taylor & Francis, **1**(1), pp. 85–95.

Description, G. and Diagram, C., F. Table, *54LS74 DM54LS74A DM74LS74A Dual Positive – Edge – Triggered D Flip-Flops with Preset Clear and Complementary*.

Desrayaud, G., Fichera, A. and Lauriat, G. (2007) 'Natural convection air-cooling of a substrate-mounted protruding heat source in a stack of parallel boards', *International Journal of Heat and Fluid Flow*. Elsevier, **28**(3), pp. 469–482.

Du, S.-Q., Bilgen, E. and Vasseur, P. (1998) 'Mixed convection heat transfer in open-ended channels with protruding heaters', *Heat and Mass Transfer*. Springer, **34**(4), pp. 263–270.

Durgam, S., Venkateshan, S.P. and Sundararajan, T. (2018) 'A novel concept of discrete heat source array with dummy components cooled by forced convection in a vertical channel', *Applied Thermal Engineering*. Elsevier, **129**, pp. 979–994.

Durgam, S., Venkateshan, S.P. and Sundararajan, T. (2019) 'Conjugate forced convection from heat sources on substrates of different thermal conductivity', *Journal of Thermophysics and Heat Transfer*. American Institute of Aeronautics and Astronautics, pp. 1–13.

Ekbote, A., Karvinkoppa, M., Bhojwani, V. and Patil, N. (2020) 'Comprehensive study on smart cooling techniques used for batteries', *E3S Web of Conferences*. EDP Sciences, **170**, p. 01028.

El Qarnia, H., Draoui, A. and Lakhal, E.K. (2013) 'Computation of melting with natural convection inside a rectangular enclosure heated by discrete protruding heat sources', *Applied Mathematical Modelling*. Elsevier, **37**(6), pp. 3968–3981.

Elmozughi, A.F., Solomon, L., Oztekin, A. and Neti, S. (2014) 'Encapsulated phase change material for high-temperature thermal energy storage–Heat transfer analysis', *International Journal of Heat and Mass Transfer*. Elsevier, **78**, pp. 1135–1144.

Faraji, M. and El Qarnia, H. (2010) 'Numerical study of melting in an enclosure with discrete protruding heat sources', *Applied Mathematical Modelling*. Elsevier, **34**(5), pp. 1258–1275.

Florio, L.A. and Harnoy, A. (2007) 'Use of a vibrating plate to enhance natural convection cooling of a discrete heat source in a vertical channel', *Applied Thermal Engineering*. Elsevier, **27**(13), pp. 2276–2293.

Fujii, M., Gima, S., Tomimura, T. and Zhang, X. (1996) 'Natural convection to air from an array of vertical parallel plates with discrete and protruding heat sources', *International Journal of Heat and Fluid Flow*. Elsevier, **17**(5), pp. 483–490.

Garcia, F., Treviño, C., Lizardi, J. and Martínez-Suástegui, L. (2019) 'Numerical study of buoyancy and inclination effects on transient mixed convection in a channel with two facing cavities with discrete heating', *International Journal of Mechanical Sciences*. Elsevier, **155**, pp. 295–314.

Gasia, J., Maldonado, J.M., Galati, F., De Simone, M. and Cabeza, L.F. (2019) 'Experimental evaluation of the use of fins and metal wool as heat transfer enhancement techniques in a latent heat thermal energy storage system', *Energy Conversion and Management*. Elsevier, **184**, pp. 530–538.

Gharbi, S., Harmand, S. and Jabrallah, S.B. (2018) 'Parametric study on thermal performance of PCM heat sink used for electronic cooling', in *Exergy for a Better Environment and Improved Sustainability 1*. Springer, pp. 243–256.

Guimarães, P.M. and Menon, G.J. (2003). 'Mixed convection in an inclined channel with a discrete heat source', *Mecánica Computacional*. ceride.gov.ar, pp.1667–1681.

Guimarães, P.M. and Menon, G.J. (2008) 'Combined free and forced convection in an inclined channel with discrete heat sources', *International Communications in Heat and Mass Transfer*. Elsevier, **35**(10), pp. 1267–1274.

Habib, M.A., Said, S.A.M. and Ayinde, T. (2014) 'Characteristics of natural convection heat transfer in an array of discrete heat sources', *Experimental Heat Transfer*. Taylor & Francis, **27**(1), pp. 91–111.

Hamza, F. and Mustapha, F. (2019) 'Effect of Nanoparticles Insertion on Heat Storage Efficiency in a Phase Change Material', in *2019 International Conference on Wireless Technologies, Embedded and Intelligent Systems (WITS)*, pp. 1–5.

Hasan, A., Hejase, H., Abdelbaqi, S., Assi, A. and Hamdan, M. (2016) 'Comparative effectiveness of different phase change materials to improve cooling performance of heat sinks for electronic devices', *Applied Sciences*. Multidisciplinary Digital Publishing Institute, **6**(9), p. 226.

Hasan, M.I. and Tbena, H.L. (2018a) 'Enhancing the cooling performance of micro pin fin heat sink by using the phase change materials with different configurations', in *2018 International Conference on Advance of Sustainable Engineering and its Application (ICASEA)*, pp. 205–209.

Hasan, M.I. and Tbena, H.L. (2018b) 'Using of phase change materials to enhance the thermal performance of microchannel heat sink', *Engineering Science and Technology, an International Journal*. Elsevier, **21**(3), pp. 517–526.

Instruments, T. and Slvs, I. (2002) '*TPS3510, TPS3511: PC Power Supply Supervisors (Rev. A)*'.

Janarthanan, B. and Sagadevan, S. (2015) 'Thermal energy storage using phase change materials and their applications: A review', *International Journal of ChemTech Research*. academia, **8**, pp. 250–256.

Kahwaji, S., Johnson, M.B., Kheirabadi, A.C., Groulx, D. and White, M.A. (2018) 'A comprehensive study of properties of paraffin phase change materials for solar thermal energy storage and thermal management applications', *Energy*. Elsevier, **162**, pp. 1169–1182.

Kalbasi, R., Afrand, M., Alsarraf, J. and Tran, M.D. (2019) 'Studies on optimum fins number in PCM-based heat sinks', *Energy*. Elsevier, **171**, pp. 1088–1099.

Kandasamy, R., Wang, X.-Q. and Mujumdar, A.S. (2008) 'Transient cooling of electronics using phase change material (PCM)-based heat sinks', *Applied Thermal Engineering*. Elsevier, **28**(8–9), pp. 1047–1057.

Karami, R. and Kamkari, B. (2019) 'Investigation of the effect of inclination angle on the melting enhancement of phase change material in finned latent heat thermal storage units', *Applied Thermal Engineering*. Elsevier, 146, pp. 45–60.

Kargar, A., Ghasemi, B. and Aminossadati, S.M. (2011) 'An artificial neural network approach to cooling analysis of electronic components in enclosures filled with nanofluids', *Journal of Electronic Packaging*. American Society of Mechanical Engineers, 133(1), p. 11010.

Karvinkoppa, M.V. and Hotta, T.K. (2017). 'Numerical investigation of natural and mixed convection heat transfer on optimal distribution of discrete heat sources mounted on a substrate', *IOP Conference Series: Materials Science and Engineering*. IOP Publishing, 263(6), p. 062066.

Karvinkoppa, M.V. and Hotta, T.K. (2019) 'Transient analysis of phase change material for the cooling of discrete heat sources under mixed convection', *International Society for Energy, Environment and Sustainability*, 8, 1–6.

Khademi, A., Darbandi, M., Behshad Shafii, M. and Schneider, G. (2019) 'Numerical simulation of phase change materials to predict the energy storage process accurately', in *AIAA Propulsion and Energy 2019 Forum*, p. 4225.

Koca, A., Oztop, H.F. and Varol, Y. (2008) 'Natural convection analysis for both protruding and flush-mounted heaters located in triangular enclosure', *Proceedings of the Institution of Mechanical Engineers, Part C: Journal of Mechanical Engineering Science*. SAGE Publications Sage UK: London, England, 222(7), pp. 1203–1214.

Korichi, A. and Laouche, N. (2019) 'Pulsed flow in a vertical channel with discrete heat sources', *SSRN*. 3372913.

Kumar, V.G. and Phaneendra, K. (2019) 'Optimization of temperature of a 3D duct with the position of heat sources under mixed convection', in *Numerical Heat Transfer and Fluid Flow*. Springer, pp. 275–284.

Kurhade, A., Talele, V., Rao, T.V., Chandak, A. and Mathew, V.K. (2021a) 'Computational study of PCM cooling for electronic circuit of smart-phone', *Materials Today: Proceedings*. Elsevier, 47, pp. 3171–3176.

Kurhade, A.S., Rao, T. V., Mathew, V.K., and Patil, N.G. (2021b). 'Effect of thermal conductivity of substrate board for temperature control of electronic components: A numerical study', *International Journal of Modern Physics C (IJMPC)*. World Scientific Publication, 32(10), pp. 1–12.

Ling, Z., Chen, J., Fang, X., Zhang, Z., Xu, T., Gao, X. and Wang, S. (2014) 'Experimental and numerical investigation of the application of phase change materials in a simulative power batteries thermal management system', *Applied Energy*. Elsevier, 121, pp. 104–113.

Liu, Y., Phan-Thien, N., Kemp, R. and Luo, X.L. (1997) 'Three-dimensional coupled conduction-convection problem for three chips mounted on a substrate in an enclosure', *Numerical Heat Transfer, Part A: Applications*. Taylor & Francis, 32(2), pp. 149–167.

Loganathan, A. and Mani, I. (2018) 'A fuzzy-based hybrid multi-criteria decision-making methodology for phase change material selection in electronics cooling system', *Ain Shams Engineering Journal*. Elsevier, 9(4), pp. 2943–2950.

Madadi, R.R. and Balaji, C. (2008) 'Optimization of the location of multiple discrete heat sources in a ventilated cavity using artificial neural networks and micro genetic algorithm', *International Journal of Heat and Mass Transfer.* Elsevier, 51(9–10), pp. 2299–2312.

Mahdi, M.S. (2013) 'Natural convection in a vertical rectangular enclosure', *Kirkuk University Journal for Scientific Studies.* Kirkuk University, 8(4), pp. 1–16.

Mahmoud, S., Tang, A., Toh, C., Raya, A. L. D. and Soo, S. L. (2013) 'Experimental investigation of inserts configurations and PCM type on the thermal performance of PCM based heat sinks', *Applied Energy.* Elsevier, 112, 1349–1356.

Mathew, V.K. and Hotta, T.K. (2018). 'Numerical investigation on optimal arrangement of IC chips mounted on a SMPS board cooled under mixed convection', *Thermal Science and Engineering Progress.* Elsevier, 7, 221–229.

Mathew, V.K. and Hotta, T.K. (2020). 'Experiment and numerical investigation on optimal distribution of discrete ICs for different orientation of substrate board', *International Journal of Ambient Energy.* Taylor & Francis, 1–8.

Mathew, V.K. and Hotta, T.K. (2021a) 'Experimental investigation of substrate board orientation effect on the optimal distribution of IC chips under forced convection', *Experimental Heat Transfer.* Taylor & Francis, 34(6), 564–585.

Mathew, V.K. and Hotta, T.K. (2021b). 'Performance enhancement of high heat generating IC chips using paraffin wax based mini-channels – A combined experimental and numerical approach' *International Journal of Thermal Sciences.* Elsevier, 164, p. 106865.

Modulator, P.W., C. Circuits, Pulse Width Modulator Control Circuits SG3525A.

Narasimham, G. (2010) 'Natural convection from discrete heat sources in enclosures: An overview', *Vivechan International Journal of Research.* IMSEC, 1, pp. 63–78.

Nardini, G. and Paroncini, M. (2012) 'Heat transfer experiment on natural convection in a square cavity with discrete sources', *Heat and Mass Transfer.* Springer, 48(11), pp. 1855–1865.

Order A J Theory Ordered Sets. (1996) *'Its Application 1–8.'*

Ozsunar, A., Arcaklioglu, E. and Dur, F.N. (2009) 'The prediction of maximum temperature for single chips' cooling using artificial neural networks', *Heat and Mass Transfer.* Springer, 45(4), pp. 443–450.

Ozsunar, A., Baskaya, S. and Sivrioglu, M. (2001) 'Numerical analysis of Grashof number, Reynolds number and inclination effects on mixed convection heat transfer in rectangular channels', *International Communications in Heat and Mass Transfer.* Elsevier, 28(7), pp. 985–994.

Panthalookaran, V. (2010) 'CFD-assisted optimization of chimney like flows to cool an electronic device', *Journal of Electronic Packaging.* American Society of Mechanical Engineers, 132(3), p. 31007.

Premachandran, B. and Balaji, C. (2005) 'Mixed convection heat transfer from a horizontal channel with protruding heat sources', *Heat and Mass Transfer.* Springer, 41(6), pp. 510–518.

Queipo, N.V. and Gil, G.F. (1999) 'Multiobjective Optimal Placement of Convectively and Conductively Cooled Electronic Components on Printed Wiring Boards'.

Rabie, R., Emam, M., Ookawara, S. and Ahmed, M. (2019) 'Thermal management of concentrator photovoltaic systems using new configurations of phase change material heat sinks', *Solar Energy*. Elsevier, **183**, pp. 632–652.

Rathod, M. K. and Banerjee, J. (2014) 'Experimental investigations on latent heat storage unit using paraffin wax as phase change material', *Experimental Heat Transfer*. Taylor & Francis, **27**(1), pp. 40–55.

Rehman, T. and Ali, H. M. (2018) 'Experimental investigation on paraffin wax integrated with copper foam based heat sinks for electronic components thermal cooling', *International Communications in Heat and Mass Transfer*. Elsevier, **98**, 155–162.

Roache, P.J. (2013). 'Perspective: A method for uniform reporting of grid refinement studies'. *Journal of Fluids Engineering*. United States: N. p, **116**(3), pp. 405–413.

Roy, N. C., Hossain, M. A. and Gorla, R. S. R. (2020) 'Natural convection in a cavity with trapezoidal heat sources mounted on a square cylinder', *SN Applied Sciences*. Springer, **2**, 1–11.

Saha, S., Nayak, K.C., Srinivasan, K. and Dutta, P. (2006) 'Cooling of electronics using phase change materials and thermal conductivity enhancers', in *18th National & 7th ISHMTASME Heat Mass Transfer Conference January*. ISHMT-ASME HMT-2006-C101, pp. 4–6.

Saha, S.K. and Dutta, P. (2010) 'Cooling of electronics with phase change materials', in *AIP Conference Proceedings*, pp. 31–36.

Salunkhe, P.B. and Shembekar, P.S. (2012) 'A review on the effect of phase change material encapsulation on the thermal performance of a system', *Renewable and Sustainable Energy Reviews*. Elsevier, **16**(8), pp. 5603–5616.

Senthil, R. and Cheralathan, M. (2016) 'Natural heat transfer enhancement methods in phase change material based thermal energy storage', *International Journal of ChemTech Research*. SPHINXSAI, **9**(5), pp. 563–570.

Sharma, A., Tyagi, V.V., Chen, C.R. and Buddhi, D. (2009) 'Review on thermal energy storage with phase change materials and applications', *Renew and Sustainable Energy Reviews*. Elsevier, **13**(2), pp. 318–345.

Solomon, A. (1981) 'A note on the Stefan number in slab melting and solidification', *Letters in Heat and Mass Transfer*. Elsevier, **8**(3), pp. 229–235.

Song, L., Zhang, H. and Yang, C. (2019) 'Thermal analysis of conjugated cooling configurations using phase change material and liquid cooling techniques for a battery module', *International Journal of Heat and Mass Transfer*. Elsevier, **133**, pp. 827–841.

Sudhakar, T.V.V., Balaji, C. and Venkateshan, S.P. (2010a) 'A heuristic approach to optimal arrangement of multiple heat sources under conjugate natural convection', *International Journal of Heat and Mass Transfer*. Elsevier, **53**(1–3), pp. 431–444.

Sudhakar, T.V.V., Shori, A., Balaji, C. and Venkateshan, S.P. (2010b) 'Optimal heat distribution among discrete protruding heat sources in a vertical duct: A combined numerical and experimental study', *Journal of Heat Transfer*. American Society of Mechanical Engineers, **132**(1), p. 11401.

Talele, V., Thorat, P., Gokhale, Y.P. and Mathew, V.K. (2021). 'Phase change material based passive battery thermal management system to predict delay effect', *Journal of Energy Storage*. Elsevier, **44**, 103482.

Talukdar, D., Li, C.G. and Tsubokura, M. (2019) '*Investigation of compressible laminar natural-convection for a staggered and symmetric arrangement of discrete heat sources in an open-ended vertical channel*', Numerical Heat Transfer, Part A: Applications. Taylor & Francis, **76**, 115–138.

Texas Instruments. (2014) '*LMx58 – N Low – Power, Dual – Operational Amplifiers*'.

Thomas, J., Srivatsa, P.V.S.S., Krishnan, S.R. and Baby, R. (2016) 'Thermal performance evaluation of a phase change material based heat sink: A numerical study'. *Procedia Technology*. Elsevier, **25**, pp. 1182–1190.

Totten, G.E., Westbrook, S.R. and Shah, R.J. (2003) '*Fuels and Lubricants Handbook*', ASTM International, 593–594.

Tou, S.K.W. and Zhang, X.F. (2003) 'Three-dimensional numerical simulation of natural convection in an inclined liquid-filled enclosure with an array of discrete heaters', *International Journal of Heat and Mass Transfer*. Elsevier, **46**(1), pp. 127–138.

Usman, H., Ali, H.M., Arshad, A., Ashraf, M.J., Khushnood, S., Janjua, M.M. and Kazi, S.N. (2018) 'An experimental study of PCM based finned and unfinned heat sinks for passive cooling of electronics', *Heat and Mass Transfer*. Springer, **54**(12), pp. 3587–3598.

Venkateshan, S.P. (2004) '*First Course in Heat Transfer*', Ane Books, New Delhi, India.

Venkateshan, S.P. (2008) '*Mechanical Measurements*', Ane Books, New Delhi, India.

Vishay, High-Speed Optocoupler, 1 MBd. (2002) '*Photodiode with Transistor Output*'.

Voller, V.R. and Prakash, C. (1987) 'A fixed grid numerical modelling methodology for convection-diffusion mushy region phase-change problems'. *International Journal of Heat and Mass Transfer*. Elsevier, **30**(8), pp. 1709–1719.

Xie, J., Lee, H.M. and Xiang, J. (2019) 'Numerical study of thermally optimized metal structures in a phase change material (PCM) enclosure', *Applied Thermal Engineering*. Elsevier, **148**, pp. 825–837.

Yavari, F., Fard, H.R., Pashayi, K., Rafiee, M.A., Zamiri, A., Yu, Z., Ozisik, R., Borca-Tasciuc, T. and Koratkar N. (2011) 'Enhanced thermal conductivity in a nanostructured phase change composite due to low concentration graphene additives', *The Journal of Physical Chemistry C*. ACS Publications, **115**(17), pp. 8753–8758.

Ye, W.-B., Zhu, D.-S. and Wang, N. (2011) 'Numerical simulation on phase-change thermal storage/release in a plate-fin unit', *Applied Thermal Engineering*. Elsevier, **31**(17–18), pp. 3871–3884.

Zalba, B., Marın, J.M., Cabeza, L.F. and Mehling, H. (2003) 'Review on thermal energy storage with phase change: materials, heat transfer analysis and applications', *Applied Thermal Engineering*. Elsevier, **23**(3), pp. 251–283.

Zarma, I., Ahmed, M. and Ookawara, S. (2019) 'Enhancing the performance of concentrator photovoltaic systems using nanoparticle-phase change material heat sinks', *Energy Conversion and Management*. Elsevier, **179**, pp. 229–242.

Zeng, J.L., Cao, Z., Yang, D.W., Sun, L.X. and Zhang, L. (2010) 'Thermal conductivity enhancement of ag nanowires on an organic phase change material', *Journal of Thermal Analysis and Calorimetry*. Springer, **101**(1), pp. 385–389.

Zhou, J., Chen, Z., Liu, D. and Li, J. (2001) 'Experimental study on melting in a rectangular enclosure heated below with discrete heat sources', *Journal of Thermal Science*. Springer, **10**(3), pp. 254–259.

Appendix A

MATLAB® program for generating all the possible configurations for the arrangement of seven non-identical rectangular IC chips on a substrate board

```
clc;
clear;
Wkg_area=101.6; %Working area size
G=[7.60 10.45;4 5;7.60 10.5;4 5];
e=size(G,1);
Sum_A=0;
Fx=[86.84 68.71 50.02];%X Centroid of fixed heat
sources
Fy=[49.93 114.43 119.175];%Y Centroid of fixed heat
sources
for f=1:1:e
    Sum_A=Sum_A+G(f,1)*G(f,2);
end
comb_count=size(C,1);
t=0;
k=32;
for l=1:1:comb_count
    Q=perms(C(l,:));
    r=size(Q,1);
        for s=1:1:r
            t=t+1;
            Z(t,:)=Q(s,:);
            sum_Ax=0;
            sum_Ay=0;
                for m=1:1:e
                    for r=1:1:k
                        if(Z(t,m)== T(r,1))
                            x(m) = T(r,2)+G(m,1)/2;
```

```
                                  y(m)  =  T(r,3)+G(m,2)/2;
                                  sum_Ax = sum_Ax +
x(m)*G(m,1)*G(m,2);
                                  sum Ay = sum_Ay +
y(m)*G(m,1)*G(m,2);
                         break;
                         end
                    end
               end
     xc = (sum_Ax+18431.36106)/(Sum_A+282.5634);
     yc = (sum_Ay+27946.24659)/(Sum_A+282.5634);
     di_square=0;
     di_sqre_fixed=0;
               for m=1:1:e
                    di_square = di_square +
(x(m)-xc)^2 + (y(m)-yc)^2;
                    end
                    for f=1:3
                         di_sqre_fixed=di_sqre_
fixed+(Fx(f)-xc)^2 + (Fy(f)-yc)^2;
                    end
Lamda(t,1)=(di_square+di_sqre_fixed)/(Wkg_area^2 +
yc^2);
          end
end
Y=[Lamda Z];
S=sortrows(Y,1);
a1=S(1:100000,1:5);
a2=S(100001:200000,1:5);
a3=S(200001:300000,1:5);
a4=S(300001:400000,1:5);
a5=S(400001:500000,1:5);
a6=S(500001:600000,1:5);
a7=S(600001:700000,1:5);
a8=S(700001:800000,1:5);
a9=S(800001:863040,1:5);
```

Appendix B

Calculation of mixed convection considered for the numerical study

Mixed convection is characterized by Richardson number (Ri), which is given as

$$1 = \frac{Gr}{\text{Re}^2} = Ri$$

$$Gr = \frac{g\beta\Delta T_{ref}H^3}{v^2}, \text{Re} = \frac{uL_h}{v}$$

$$\beta = \frac{1}{T_{mean}} = \frac{1}{(55+273)} = 3.04878 \times 10^{-3}\,\text{K}^{-1},$$

$$L_h = \frac{4A}{P} = \frac{4 \times (0.01965 \times 0.0065)}{(0.01965 \times 0.0065)} = 9.7686 \times 10^{-3}$$

$$\Delta T_{ref} = \frac{qL}{K_s} = \frac{54600 \times 0.01965}{0.35} = 3065.4\text{K}$$

$$1 = \frac{9.81 \times 3.04878 \times 10^{-3} \times 3065.4 \times 0.1^3}{(u \times 9.7686 \times 10^{-3})^2}$$

$$u = 30.996 \approx 31 m/s$$

Appendix C

Sample calculation for non-dimensional temperature (θ) and Fourier number (Fo)

$$\Delta T_{ref} = \frac{qL}{K_s} = \frac{54600 * 0.01965}{0.24} = 4470.375$$

$$\theta = \frac{(T_{max} - T_\infty)}{\Delta T_{ref}} = \frac{(61.48 - 24)}{4470.375} = 0.008160998$$

$$Fo = \frac{\alpha_{pcm} t}{L_c^2} = \frac{(1.21423e - 07)}{0.01965^2} = 0.047170247$$

Appendix D

Sample calculation for Uncertainty in the Heat supplied to the IC chip

$$Q = VI$$

Where Q is the power input W, V is the supply voltage V, and I is the current corresponding to the supply voltage A

The uncertainty in the voltage, ΔV is 15 ± 0.05 V of the full scale

The uncertainty in the current, ΔI is 1 ± 0.002 A of the full scale

So, the uncertainty in the power input, ΔQ is calculated using the formula given in Eqn. 9, and is given below

$$Q = \pm \sqrt{\left(\frac{\partial Q}{\partial V} \times \Delta V\right)^2 + \left(\frac{\partial Q}{\partial I} \times \Delta I\right)^2}$$

$$Q = \pm \sqrt{(I \times \Delta V)^2 + (V \times \Delta I)^2}$$

$$Q = \pm \sqrt{(1 \times 0.05)^2 + (15 \times 0.002)^2}$$

$$Q = \pm 0.0583 \ W$$

In term of percentage, it is

$$\frac{\Delta Q}{Q} \times 100 = \frac{\pm 0.0583}{15 \times 1} \times 100 = 0.388\%$$

Index